HOME
WIND POWER

United States Department of Energy

Garden Way Publishing

Charlotte, Vermont 05445

This material was originally published in 1978 under the title *Wind Power for Farms, Homes and Small Industry* by the United States Department of Energy.

Garden Way Publishing Co. is grateful for assistance, both large and small, provided by Enertech of Norwich, Vermont; the Wind Systems Program, Rocky Flats Plant, Golden, Colorado; the Solar Energy Research Institute, Golden, Colorado; the United States Department of Energy, Washington, D.C.; and the many manufacturers whose products are represented in the catalog section of this book.

Second printing, August 1981

Library of Congress Cataloging in Publication Data
Main entry under title:

Home wind power.

Originally published: Wind power for farms, homes, and small industry. Washington, D.C.: U.S. Dept. of Energy, 1978.
Bibliography: p.
Includes index.
1. Wind power. 2. Dwellings – Power supply.
I. United States. Dept of Energy.
TK1541.H65 1980   621.31′2136   81-2799
ISBN 0-88266-252-X (pbk.)   AACR2

Contents

CHAPTER 1

1   Introduction – Is the Wind a Practical Source of Power for Me?

A summary of steps to assist in your decision-making process

CHAPTER 2

8   Wind Power – How It Works

Basic wind turbine aerodynamics • Work, energy, and power • Wind power • Calculating power

CHAPTER 3

18   Wind Behavior and Site Selection

Wind power variations with time • Wind and energy roses • Wind power distribution • Local power variations with height and terrain • Estimating your wind power

CHAPTER 4

46   Power and Energy Requirements

Electric load estimation • Mechanical load estimation • Energy storage

CHAPTER 5

63   The Components of a Wind Energy Conversion System

Wind system performance: comparing different types of wind machines • Wind system power and energy calculations • Wind machine rotor construction • Rotor control • Electric power generation • Generators and regulators • Energy storage: batteries, pumped water storage, hot water or hot air storage, flywheels, synchronous inversion • Wind system towers • Other equipment

iii

CHAPTER 6

100   Selecting Your Wind Energy Conversion System and Figuring the Cost of Its Power

Thinking out your most appropriate system • Comparison of power costs using two different wind turbines • Gathering the facts for your economic analysis—total installed cost, expected system life, total energy yield, annual costs, resale value, savings, cost of inflation and future value

CHAPTER 7

113   Possible Legal Hurdles

Deed restrictions • Wind rights and "negative easements" • Obtaining a building permit: zoning, other regulations, and building codes • Sharing, buying, and selling power: relations with neighbors and your utility • Warranty, liability, and insurance

CHAPTER 8

124   This Book and Your Wind System—Examples and Other Wind Energy Information

132   Glossary

135   Additional References on Wind Energy

APPENDIX 1

136   Wind Power in the United States and Canada

APPENDIX 2

154   Owning a Wind System

APPENDIX 3

162   English-Metric Units Conversion Table

APPENDIX 4

165   Additional Wind Behavior and Site Selection Information

179   Wind Power Catalog

201   Index

CHAPTER 1

Introduction – Is the Wind a Practical Source of Power for Me?

Wind systems have caught the public imagination. The idea of installing a machine that produces power out of thin air, allows its owner to thumb his nose at utility bills and turns a home into an energy-self-sufficient castle is immensely appealing. Those who, in increasing numbers, have sought to make this idea a reality have found that harnessing the wind is usually neither as inexpensive nor as easy as it sounds. Unless you build your own, the initial cost of even a small wind turbine can be high. In most cases, utility power is still needed during windless periods and times of high power demand. And even with storage batteries, many wind system owners have found that wind power can meet only part of their energy requirement.

But despite these drawbacks and limitations, wind-machine ownership can still be a satisfying experience. If your power costs are high, if you need mechanical or electrical power in a remote location away from existing utility lines, if you live in an area with documented high annual average winds, or if the use of alternate energy sources makes practical and philosophical sense to you, installing a wind turbine may offer definite economic and operational, as well as emotional, rewards.

This publication is intended to provide you with a basis for determining the practicality of wind energy for your particular situation. Whether or not it is a practical solution depends upon your specific energy needs and a variety of other considerations. If you decide that the wind is a practical energy source for you, other decisions are required, such as the type and amount of equipment you will need and whether to buy a wind energy conversion system (WECS) or build your own. This book will help you make these decisions.

To decide if wind energy is practical for you, you will want to determine your energy requirements, your available wind energy resource and the equipment needed to convert and use the available energy. You will have to consider the cost of energy obtained from a wind system and decide if the wind is a practical source of power for you. This publication will provide you with the basic information or methods you need to make these decisions. You may already be familiar with material in some of the chapters. In that case, use this introduction as a guide to determining the chapters in which you would like to concentrate your efforts.

1

STEP-BY-STEP DECISION MAKING

Figure 1–1 shows the steps necessary in your decision-making process. Each step in the table is discussed briefly in this introductory chapter; however, subsequent chapters provide more detailed information about each step. With the aid of this book, you may choose to proceed on your own through the entire decision-making process. Alternatives are to hire a consultant, or perhaps to seek advice from a manufacturer or distributor of wind machines.

The steps listed in Figure 1–1 are sequential. The results determined at any given step, however, may negate results already obtained from previous steps. You then may want to repeat an earlier step in search of a different answer that will satisfy the new information.

Figure 1-1: Steps for determining the practicality of a wind system.

- Evaluate the legal and environmental impacts (Chapter 7, Appendix 2).

- Evaluate your energy requirements (Chapter 4).

- Evaluate the wind resource at your proposed location (Chapter 3).

- Select system components (Chapter 5, 6).

- Evaluate cost of the system (Chapter 6).

- Re-evaluate energy requirements and legal and environmental impacts if necessary.

- Evaluate alternatives in buying, installing, and owning a wind system.

Evaluate Legal and Environmental Impacts

If your site is in a remote rural area, you may want to go directly to Step 2. But if it is in an area with strict building or zoning codes, check the applicability of these codes to a wind turbine. Examples of possible restrictions include structure height and distance from property lines and roads. There also are other considerations to keep in mind as you proceed through the decision-making steps outlined in Figure 1–1. There may be environmental issues such as visual impact and noise from your wind system. Social issues may arise such as getting along with your neighbors while your wind machine blocks their view of the sunset. You may be confronted with a "wind rights" problem (your neighbor erects a tall building upwind of your shiny new machine and renders it an idle art form in your yard). You must think each possible issue through and resolve whether or not to proceed further in your WECS plan.

Evaluate Energy Requirements

Step 2 in Figure 1–1 calls for calculating your total energy requirements. One method is to look at your various monthly electric bills. These will tell how much you have used in the past. Try to relate this information to how you expect to use energy after you switch to a wind machine. Another way to assess your needs is to calculate the monthly energy demands (expressed in kilowatt-hours or kWh) by adding up the energy required for most appliances and farm equipment in current use.

For pumping water, it is necessary to calculate the needed horsepower and the hours of pumping needed. This gives you horsepower-hours which, like kilowatt-hours, can be used as an energy assessment. Methods for making

Figure 1-2: A remote-site wind system atop Mount Washington in New Hampshire.

these calculations are given in Chapter 4. In some cases, you can simply determine gallons per hour of water you must pump and the total height you must raise the water (from well bottom to tank top, for example). Some water-pumper manufacturers present performance data for their wind machines in terms of gallons per hour, height of water lift, and wind speed.

Evaluate Wind Resource

The third step calls for an evaluation of the wind resource at your proposed site. This step requires more than holding up a wet finger or seeking the advice of neighbors. Most people tend to overestimate wind speeds. The amount of time and money you spend on this task is directed by your need for accuracy and by the size and importance of your wind system (from the standpoint of safety and energy production). Your options are hiring a consultant to perform a wind site survey or doing the survey yourself.

Techniques used in a site survey have varying degrees of effectiveness. The crudest, of course, is the wet finger to find intuitively places to install a wind machine. Another approach is the use of the Beaufort Scale shown in Chapter 3. That table relates wind speed to the movement of trees and to other effects of the wind. You might contact a weather service for climatic data for your area or perhaps a local airport for its average wind speed. It is unlikely, however, that either source will have information pertinent to the exact area you have in mind. Average monthly wind power calculations for over 700 locations in the United States and Canada are included in Appendix 1. The most effective approach is using instruments—the placement of wind speed and wind direction equipment—to get actual site readings.

During your intuitive exploration and/or site instrumentation, you should have an awareness of the effects on wind speed by man-made and natural blockages. Examples are buildings, trees, and other features of the terrain. As shown in Figure 3-14, these obstructions interfere with normal wind speeds at given heights and thus will influence your eventual tower height.

In the field, you might measure a wind speed of 20 mph at about 30 meters high (98.5 feet, which is very high for a tower intended for a home or farm wind system). Back in the city, however, you may have to go as high as 200 meters to find the same wind speed as at 30 meters in the country. This is due to the interference that buildings have on air flow. It's somewhat better in the suburbs, but you can see that open country causes the least resistance to wind, and that higher towers will raise your machine up into strong winds. Much of Chapter 3 is devoted to the effects of obstructions on air flow.

Strong winds have a critical effect on wind machines: *a doubling of wind speed results in eight times more power available to your wind machine*. This means that a location with an annual average wind speed of 12.6 mph offers twice the energy available compared to a site with a 10 mph average. This phenomenon is discussed more in Chapter 2.

Calculating an average wind speed per year based on limited observations can be difficult at best. Winds vary considerably during the year when viewed on a monthly basis. This variation is illustrated in Figure 3-3 for three different locations. Your site may have strong winds in winter and weak winds in summer. A nearby site may have little or no wind at all.

Select System and Components

After analyzing available data to this point, you should know your energy requirements and energy resources (monthly and annual wind speed). Now is the time to match this information to determine the windmill size that will be needed. One means of doing this is shown in Figure 5-17B. As an example of how that illustration can be used, suppose you determine that 500 kilowatt-hours of energy will be required. By means of a few simple calculations (described in Chapter 5), it is resolved that 2000 watts of power would be required from the wind generator when the wind blows at its average speed. Suppose now that your site analysis indicated an average wind speed of 10 mph. Referring to Figure 5-17B, it is noted that no curve for windmill diameters crosses the intersection of 10 mph and 2000 watts. If another curve were to be drawn across this point, it probably would indicate a 40- or 50-foot diameter machine. If you think that this size windmill is too large for you, your budget, neighbors, or site, you have two alternatives: reduce your energy needs or find a site with a greater average windspeed.

As another example, assume that you reduced the energy needs just mentioned to 500 watts. On Figure 5-17B, a curve crosses very close to the intersection of 500 watts and 10 mph. That curve indicates a wind machine 20 feet in diameter. You might be able to cope better with a machine this size.

These are two examples, each illustrating a different type of conclusion. The ultimate conclusion, however, is based on more than the analysis just discussed. Attention also should be given to the wind system's intended application and to the importance of size, cost, and to other factors.

Selection of the type and brand of wind system, tower, and other com-

Figure 1-3: American farm windmill.

ponents will require careful consideration. There are various types of wind machines to choose from. Figures 1–3 and 1–4, for example, show two different types of machines. They are similar, however, in having propeller-type blades mounted on a horizontal power shaft whereas the machine in Figure 5–5 has a vertical shaft. There are other shapes for wind machines, and there are several manufacturers for practically all of the proven types. More types and brands of machines are emerging continually.

The U.S. Department of Energy (headquartered in Washington, D.C.) and the American Wind Energy Association* work together closely to maintain current lists of manufacturers. The Department of Energy has a Small Wind Systems Test Center located at its Rocky Flats Plant near Golden, Colorado. Many of the different brands and types of wind machines are tested at the center in an effort to promote product improvement. Results from these tests are available to the public.†

Selection of a wind–electric machine will be followed by selection of the tower, batteries, inverters, and other devices in the case of battery systems. For a non-battery electric system, a synchronous inverter or wind furnace

---

*American Wind Energy Association, 1609 Connecticut Ave., Washington, D.C. 20009. Also see the Wind Power Catalog in this book for a partial listing and description of manufacturers and equipment suppliers.

†For information write: Darrell Dodge, Wind Energy Program, Rockwell International Energy Systems Group, P.O. Box 464, Golden, CO 80401.

Figure 1-4: A conventional wind turbine.

heating element may be selected. A water-pumper system requires the
selection of a wind machine, tower, pumps, and plumbing.

Figure 5–32 illustrates the interrelationship of components in one wind–
electric system. It shows a system in which the wind generator charges a set
of batteries where the energy is stored. The batteries then supply electric
power to various loads such as lights, motors, refrigerators, and television
sets. Included is a backup generator that charges the batteries in the event of
an extended low-wind period or during a period of extra-heavy energy usage.

Evaluate System Cost

At the same time as equipment is being selected, cost factors begin to appear. Figure 6–6 illustrates the trend usually experienced in installation planning. For small wind systems (under 30 to 40 kilowatts maximum power rating), the trend is for larger installations to be less expensive per kilowatt of
rated power.

In a complete cost analysis, you might estimate such factors as initial
costs, interest and insurance costs, maintenance costs, and energy yield over
the life of the equipment. Dividing total estimated costs (dollars) for the life
of a windmill by total energy production (kilowatt-hours) provides you with a
cost estimate of energy expressed in dollars per kilowatt-hour. You then can
compare this directly with other sources of energy or other potential wind
system installations.

Reevaluate Energy Requirements

In preparing a system plan, you may discover it sometimes is more desirable
to reduce your energy usage than to buy a larger wind machine. It is almost
always prudent to evaluate your options in the area of conservation. It also is
well to allow for future growth in energy needs, but many times you will still
find that a good system plan can be vastly improved by saving more power.

6

Three Purchase Options

After deciding what equipment is wanted, you have three options as to how the equipment will be purchased: (1) buy directly from the manufacturer; (2) buy new or used equipment from a dealer; or (3) buy used equipment from someone else. Purchasing is important enough to warrant a brief comment on at least the first two alternatives.

Buying factory-direct may seem to be a way to save time or money, and for some folks it may be; but is the factory fully equipped to come out and install your machine and maintain it? If it isn't, are you? Many wind-turbine manufacturers simply do not offer such services and you should check before you buy. If you hire someone to do such work, don't overlook the liability aspects that might be involved.

Dealers are usually organized and staffed to provide all of the services you need in addition to offering the products needed to fully equip your system. Again, it is essential to ask if the dealer provides planning, installation and maintenance services. Generally they will help you perform all of the system planning tasks discussed here.

Where you go from here depends on how helpful this introductory chapter was to you. If you need more information, use Figure 1–1 and the various chapters, as identified in the table of contents, to guide you through the book. Additional data on wind machines or manufacturers can be obtained from The American Wind Energy Association. Also, a great deal of assistance can be obtained if there is a dealer in your area.

CHAPTER 2

Wind Power – How It Works

Basic Wind Turbine Aerodynamics

Basically, wind turbines extract power from the wind when their rotors are pushed around by moving air. There are two primary ways in which the wind can accomplish this. One way is illustrated in Figure 2-1. It is a diagram of a parachute that tugs on a rope that in turn lifts a bucket of water from a well. Indeed, it is a wind machine. Important here, though, is the parachute tugging. It is caused by *drag*, which is the same force* you experience while holding your hand in the breeze outside your car while motoring along the highway. Wind is actually pushing the parachute along.

*Terms like *force*, *power*, and *energy* are discussed more technically later. Also see the Glossary.

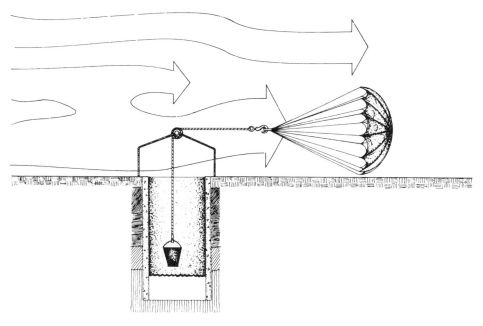

Figure 2-1: Simple wind-powered water pump.

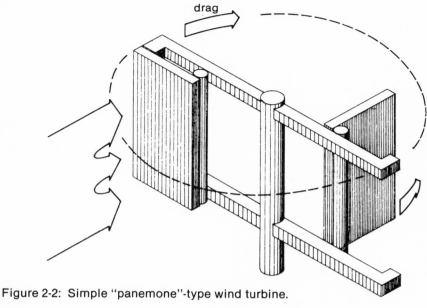

Figure 2-2: Simple "panemone"-type wind turbine.

The drag effect was used by early windmill builders to great advantage. A diagram of a simple panemone, like the machines they built, looks something like Figure 2–2. Notice that one vane is broadside to the wind. On this side of the machine, wind force (drag) will be strong. On the other side of the center shaft, the vane swings around edgewise to the wind and the drag is much less. Thus the machine turns, pivoting about the center shaft. This is how most drag machines work, although not all of them feature the pivot-mounted vanes. Elsewhere in this book, you will see other examples of this type of wind machine.

The other way in which wind can exert its force on a wind machine is by the aerodynamic action called *lift*. Lift is a force produced on airplane wings in flight (Figure 2–3). Notice that airflow around the airfoil-shaped blade tends to change direction slightly. A low-pressure area (like suction) forms over the curved side (topside) of the airfoil, and a high-pressure area (pushing upward) forms on the bottom. The result is a force upward and perpendicular to the wind direction; this makes the lift arrow point slightly forward in Figure 2–3. Drag is also produced because the wind is being slowed slightly in the process of creating lift.

Experiencing Lift and Drag

You can perform simple experiments to verify lift. Take a sheet of stiff cardboard or a wood slat about the size of a 3-by-5-inch filing card. During a trip in an automobile, at about 30 mph hold the sheet or slat out in the wind. You are now ready to experience lift and drag. Gripping one end only, hold the board with its long dimension pointing outward from the car. Hold it edgewise to the wind. Let us call the edge at the front end the *leading edge*. Point the leading edge slightly upward; you should experience a slight upward lift. Point it slightly downward; you should now experience a slight downward force (also called lift by engineers). Somewhere between upward and downward lift is an angle that produces no lift at all. See if you can find

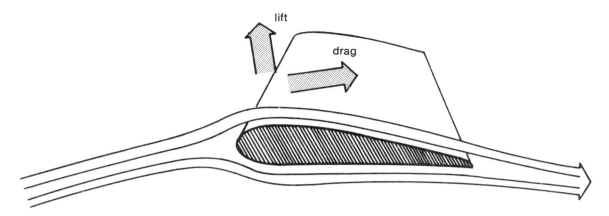

Figure 2-3: Forces acting on a wind turbine blade.

this angle of zero lift. It may not be parallel to the ground because your car bends the airflow around the windshield, fenders, and over the hood.

At the angle of zero lift, notice that a slight amount of drag is produced. Drag will tug the board aft. Now, tilt the board about 90°, leading edge up. Notice that the drag has greatly increased. You might drop the board at this point if you are moving too fast.

Now, to discover lift and drag working together, return the leading edge to the zero-lift position. Slowly rotate the board, leading edge upward. Notice lift increasing. Notice drag also increasing. Lift will increase a little faster than drag, then suddenly drop substantially while drag continues to rise. This occurs with the leading edge somewhere near 20° above the zero-lift angle. Engineers say the *wing* (your board) has *stalled.*

During this experiment, you will realize that at some particular angle lift is much greater than drag. Lift is the force used to power wind machines designed for high efficiency, and this region of highest lift with low drag is very important to windmill designers.

Wind Machines and Lift/Drag

How does a wind machine use lift as the power-producing force? Let us look at a diagram of a familiar propeller-type wind turbine (Figure 2-4). Notice that the blade performs its slight bending action on the windstream, with low-pressure and high-pressure sides similar to those shown in Figure 2-3. Lift is produced, as illustrated, generally in a direction that pushes the blade along its path.

Drag is also produced, as you might expect, and this force tries to bend the blades and slow them down in their travel around the center power shaft. In addition, the drag force tries to topple the tower that supports the wind turbine. Designers would like high lift at low drag for this type of wind machine. Figure 2-5 illustrates a typical wind generator designed for high-lift, low-drag operation.

Figure 2-6 shows a Darrieus wind turbine, sometimes called an *eggbeater* type of wind machine. Unlike the propeller type in Figure 2-5 with its power shaft pointing horizontally into the wind, the power shaft of the Darrieus is pointed across the wind. It could be horizontal as long as it pointed across the wind, but designers have found it to be more practical for the shaft to be vertical.

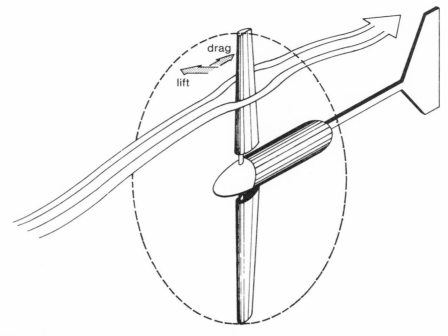

drag
lift

Figure 2-4: From lift to power—horizontal-axis wind turbine.

Figure 2-5: Propeller-type wind turbine generator.

Figure 2-6: Darrieus rotor.

11

Chances are, you found it easy to see how the propeller blade of Figure 2–4 was pulled along, but the functioning of the Darrieus blade of Figure 2–6 is less obvious. You will agree, though, that sail boats can be sailed in circles. If you sit and visualize how a sailboat has wind first on one side of the sails, then on the other as it travels around, you can begin to see how the Darrieus works.

In the propeller case (Figure 2–4), the lift force always pushes the blade along at about the same force, pivoting it about the shaft. With the Darrieus (Figure 2–6), the lift force almost always tugs the blade along its path but never with a constant force. At two areas along the path, lift is very weak. You can see that this occurs when the blade is pointed directly into the wind and directly downwind. At all other points along the blade path, lift tends to be much stronger and generally pulls the blade along its path. For this to work well, the blade must be moving along its circular path much faster than the wind is blowing. We discuss blade speed and wind speed in greater detail in Chapter 5.

Work, Energy and Power

A good understanding of work, energy, and power is not completely necessary for using this book, but in most situations it will be very helpful. For instance, does a 12-volt, 100-amp-hour battery store power or energy? The words *power* and *energy*, while often used interchangeably, have different and important meanings.

A length and force are needed to describe amounts of work. Work and velocity are needed to describe power, and power multiplied by time equals energy.

Work is performed when a force is used to lift, push, or pull some object through a distance. The amount of work done is determined by multiplying the force applied by the distance travelled (assuming the direction of the force is the same as the direction of travel). For instance, raising a 550-pound rock (250 kilograms*) one foot (0.305 meter) requires 550 foot-pounds of work.

Mechanical power is the rate at which that work is performed. That is, the force applied to an object times the velocity of the object (in the direction of the force), gives the power applied to it. For instance, if a windmill raises the 550 pounds of water at the rate of one foot per second, it is doing 550 foot-pounds of work per second, which is one horsepower (1 hp).

Electric power, as distinguished from the mechanical power just discussed, is measured in watts, kilowatts (1000 watts), and (by a power company) in megawatts (1000 kilowatts). One horsepower equals 746 watts of electric power.

If we operate a 1-hp motor for 10 hours at full capacity, 10 horsepower-hours of energy will have been consumed, assuming that the motor is 100 percent efficient. Similarly, since 1 hp equals 746 watts, then 746 × 10 = 7,460 watt-hours, or 7.46 kilowatt-hours (kWh) of electric energy will have been used, again assuming 100 percent efficiency.

More realistically, let us assume the electric energy is consumed at 50

*Common English-metric conversions of these and other measurement units that may be of use to you in your future energy considerations are given in Appendix 3.

percent efficiency. This means that half the power going into the motor is wasted. Then, to get 1 hp-hour out (7.46 kWh), we need to put twice this amount in (2 hp-hour, or 14.92 kWh).

Power is usually measured in hp (mechanical) and watts (electric), while energy is usually measured in hp-hour (mechanical) and kWh (electric). Electric energy consumption at constant power is simply power (kW) multiplied by the length of time involved (hours).

For our purposes, we can conveniently categorize energy and power as electrical or mechanical, as just described, or thermal (heat). Mechanical power is converted to electricity by a generator. A motor, for instance, converts electricity back to mechanical power. Because inefficiencies in both devices produce some heating, they usually need ventilation to remove that heat. An electric heater, of course, converts the electricity (more accurately electric power) that passes through it to heat. Heat energy is converted to mechanical energy by an engine, such as an internal combustion engine.

Mechanical power is most often generated and transferred through a rotating shaft (such as an auto drive shaft or motor shaft). Mechanical power is described as a force times a velocity. For rotating machinery, mechanical power is calculated from shaft torque times rpm.*

Mechanical energy can be either potential or kinetic energy. A weight held in your hand above the ground has potential energy. It can potentially do some damage if you drop it because the potential energy, due to its height above the ground, is continually transformed to kinetic energy as the weight speeds up. This kinetic energy is the result of speed and weight. In fact, kinetic energy increases with the square of the speed (speed times speed) while the weight is falling. That means that if the speed doubles while a weight is falling, the weight then has four times the kinetic energy.

Wind Power

The kinetic energy in the wind (energy contained in the speeding air) is proportional to the square of its velocity (just as for a falling weight). Kinetic energy in the wind is partially transformed to pressure against an object when that object is approached and air slows down. This pressure, added up over the entire object, is the total force on that object.

Power, we noted earlier, is force times velocity. This also applies to wind power. Since wind forces are proportional to the square of the velocity, wind power is proportional to wind speed cubed (multiplied by itself three times). *If the wind speed doubles, wind power goes up by a factor of eight.* This is an extremely important concept in wind power generation. To demonstrate this, consider that the blades of a conventional type of electricity-generating windmill function much like an airplane propeller (the blade shapes, however, are much different). In Figure 2–7, three propeller-driven airplanes are used to demonstrate this velocity-cubed effect.

Here, as air speed increases from 100 to 200 to 300 miles per hour (44.7, 89.4, 134.1 meters per second), the power required increases from 100 to 800 to 2700 horsepower (74.6 to 597 to 2013 kilowatts).

As you can see, increasing the speed capability of the 100-mph airplane to

---

*A useful expression relating torque and rpm to horsepower is: horsepower = 0.190 × torque × rpm ÷ 1000 where the torque is in foot-pounds.

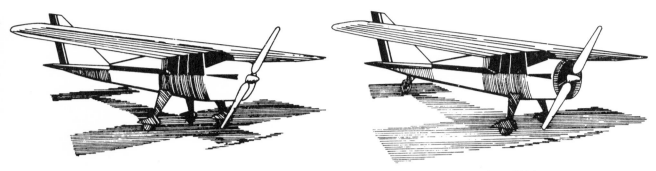

A. 100 mph from 100 hp.   B. 200 mph from 800 hp.

Figure 2-7: Power requirement increases with the cube of velocity.

200 mph means an engine change from 100 hp to 800 hp (double the speed, eight times the power). Also, to get 300 mph, we need 27 times the power, $3 \times 3 \times 3 = 27$. As shown in Figure 2–7, we have gone from a slow, good little plane to something that is bound to have to be "rocket-launched," at extreme risk to the pilot and the reputation of the structural engineers required to keep the thing together.

Incidentally, perhaps a word can be said here, drawing from the above example, about the idea of scaling up a 10-foot diameter wind turbine to, say, 100 feet in diameter. New structural problems must be considered.

Wind turbine blades take energy from the air rather than put energy in like a propeller, so when wind speed doubles, the power that can be extracted is eight times as great. When the wind speed triples, the power that can be extracted is 27 times greater. This tremendous effect of cubing the velocity can place great importance on the process of determining the best wind turbine location and emphasis upon the selection of the correct machine for your wind speeds.

Calculating Power   The power that wind turbine blades can extract from the wind is given by the expression:

$$\text{Power} = \tfrac{1}{2}\, e \times k \times A \times \varrho \times V^3$$

where:

$\times$ = multiplication
e = efficiency of the blades
k = conversion factor for units (e.g., if units on the right side are feet, pound, and seconds, and results are desired in kilowatts)
A = area swept out by the blades ($\pi \times$ blade radius^2 for conventional wind turbines)
V = wind velocity, far enough upstream so as not to be affected by the wind turbine
ϱ = Greek letter rho; equals the density of air.

The terms e, k, and ϱ need describing.

C. 300 mph from 2700 hp.

The efficiency of the blades in converting the kinetic energy in the wind to rotational power in the shaft needs some careful consideration. If all the kinetic energy in the air approaching a wind turbine were extracted by the spinning blades, the air would stop, like a car losing all its kinetic energy when it crashes into a wall. However, the air cannot stop, otherwise all the rest of the air behind it would have to spill around the rotor. Nature does not work that way. The air senses any solid object that it is approaching and moves around it, like the airflow around your automobile. When air approaches a partially solid object, such as the disc created by a spinning rotor, some of the air moves around it. The rest slows down as power is extracted by the windwheel.

Figure 2–8 illustrates how air, starting far upstream of a conventional windmill rotor, travels past the rotor. This airstream starts out being somewhat smaller than the windwheel but gradually expands to the windmill rotor size as it passes through. At this point, some of the power is taken from the wind. The power extracted by the windwheel, divided by the power in the undisturbed wind passing though a hoop the same size as the rotor, is called the *rotor power coefficient*, or more commonly, the *rotor efficiency*. Because some of the wind passes around, rather than through, the windwheel, the efficiency must be less than 100 percent.

Using laws of physics, engineers have shown that the maximum efficiency of a conventional wind system cannot exceed 59.3 percent. The same laws of nature that have been harnessed to produce our present industrialized world and send men to the moon dictate this limit. Wind system efficiencies that are claimed to be greater than this are suspect.* We have been discussing, however, horizontal-axis machines without a tip vane or surrounded with sheet metal to direct the flow. More power can be extracted from the blades if a duct is placed around the rotor, but then, if the maximum duct cross-

---

*There presently are no manufacturer's standards established for rating wind turbines. (The American Wind Energy Association is preparing standards under a federal contract. These are to be available in 1982.) Usually, wind turbines are described in terms of power, not efficiency. Occasionally, an efficiency may be stated in terms of percentage of this 59.3 percent of theoretical maximum power available. Thus, a 70 percent efficiency would mean 41.5 percent true efficiency.

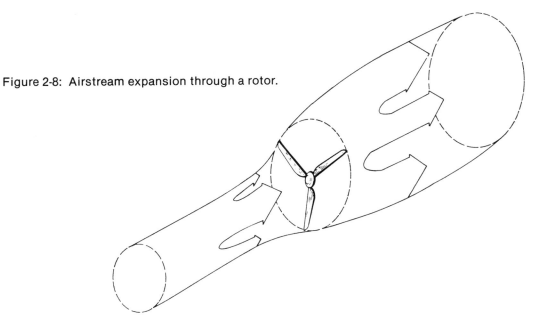

Figure 2-8: Airstream expansion through a rotor.

sectional area is used in the equation, rather than the blade rotor area, the maximum efficiency possible is still about 59.3 percent.

Well-designed blades operating at ideal conditions can extract most but not all of the 59.3 percent maximum power available. About 70 percent of this 59.3 percent is typical. Thus, a wind turbine rotor might have an advertised power coefficient, or efficiency, of $0.7 \times 0.593 = 41.5$ percent. Also, gear box, chain drive, or pulley losses, plus generator or pump losses (Chapter 5) could decrease overall wind turbine efficiency to about 30 percent. This is about the maximum coefficient possible from a conventional, well-designed wind turbine, operating at its best condition. It can be much less (see Chapter 5 and Figure 5–13 for typical wind turbine component efficiencies).

The density of air at 60°F at sea level is 0.763 pounds per cubic foot (1.22 kilogram per cubic meter). The densities at various altitudes divided by the sea level density (we will use the symbol DRA for Density Ratio at Altitude) are:

Altitude, feet	0	2,500	5,000	7,500	10,000
DRA (at 60°F)	1	0.912	0.832	0.756	0.687

The densities at various temperatures divided by the density at 60°F (we use the symbol DRT for Density Ratio at Temperature) are:

Temperature °F	0	20	40	60	80	100	120
DRT	1.130	1.083	1.040	1	0.963	0.929	0.897

To determine the true density at some particular altitude and temperature, we multiply together the appropriate DRA, DRT, and standard density of 0.0763. For example, at 100°F and 5000 feet elevation, the density $\varrho = 0.832 \times 0.929 \times 0.0763 = 0.0590$ lb/ft.[3]

The symbol k is simply a number depending on the units used for density, velocity, and area. To simplify the power equation, ½ k × (standard density) are grouped together and labeled K, so:

$$Power = K \times e \times DRA \times DRT \times A \times V^3$$

The most common units for K are:

Power	Area	Velocity	Value for k
watts	square feet	miles per hour	0.00508
watts	square feet	meters per second	0.00569
watts	square meter	miles per hour	0.0547
watts	square meter	meters per second	0.6125
watts	square feet	knots	0.00776
horsepower	square feet	miles per hour	0.00000681
horsepower	square feet	meters per second	0.0000763

The above equation has been used to calculate the curves in Figure 5–18A for DRA and DRT equal 1.0, and e = 30 percent.

As an example, a 15-foot diameter rotor is operating in a 20 mile-per-hour wind at an altitude of 500 feet and at 80°F. The windmill efficiency (including generator and transmission losses) is 30 percent. What is the power?

$$*Area = 3.14 \times 15^2 \div 4 = 176.6 \text{ square feet}$$
$$DRA = 0.832, DRT = 0.963$$
$$*Power = 0.00508 \times 0.30 \times 0.832 \times 0.963 \times 176.6 \times 20^3$$
$$= 1725 \text{ watts}$$
$$= 1.725 \text{ kilowatts}$$

Notice that Figure 5–17B shows a power of 2.2 kilowatts for this rotor at sea level and 60°F.

Finally, you want to buy power, not efficiency. If two wind turbines have the same power output in the same wind conditions, and cost and reliability are the same, it is relatively unimportant that one may be more efficient than the other.† Placing the wind turbine in the best wind location available to you and matching the power-producing velocity range of the wind turbine to your wind conditions and load is most important. This is covered in subsequent chapters.

---

*Swept area for disk-shaped windwheels is calculated from: $A = \pi \times D^2 \div 4$; π = 3.14

†Everything else being the same, the more efficient wind turbine will have a smaller rotor diameter. This could reduce its weight and cost.

CHAPTER 3

Wind Behavior and Site Selection

Two percent of all solar energy reaching the earth is converted to wind energy. Surface winds over the United States available for conversion are sufficient to supply about 30 times the country's total energy consumption. That is a huge amount of power. To reduce the scale of our thinking and to understand how winds are generated, let us first look at a few generalities about winds.

The United States and other parts of the north and south temperate zones experience a general westerly wind (Figure 3–1). Changes occur in the weather with the alternate passage of high- and low-pressure systems. These cause barometer readings to fluctuate. The various pressure systems tend to migrate from west to east and bring about wind shifts, temperature changes, rain, and other weather features.

Along with this general trend are the regional and local weather effects that are often strongly influenced by temperature differences between the air, land, and water. For example, in an area of mountains and valleys, daytime sunshine heats the mountain air, which rises and is then replaced by cooler

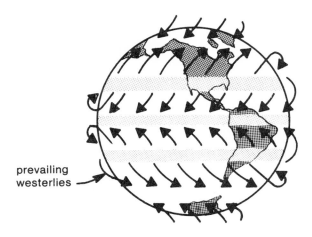

Figure 3-1: Worldwide wind circulation.

land breeze (night)   sea breeze (day)

Figure 3-2: Local wind circulation example.

air from the valley. This creates valley winds moving uphill. At night, the air cools by radiation to the night sky and moves downhill, creating a mountain breeze.

Another example of regional or local wind occurs along coastal areas. Daily temperature differences between the surfaces of the sea and land cause alternate sea and land breezes (Figure 3–2). Sea-land and valley-mountain winds are described in more detail later in this chapter. We can see that wind available for conversion is a result of both the motion in the atmosphere over a huge area and the local effects of terrain and temperature.

Understanding the characteristics of wind power variations with time is most important and is described in the first section of this chapter. The average winds in the United States are described in the second section. Local wind effects are presented in the third section. In the final section equipment and techniques for determining your best sites and their wind power potential are discussed.

Appendix 1 contains wind speed and power information for 750 stations in the United States and Southern Canada, plus other tables showing wind characteristics. For additional detailed information on the behavior of the wind and on how to perform a wind survey, see Appendix 4 and read *A Siting Handbook for Small Wind Energy Conversion Systems.* *

WIND POWER VARIATIONS WITH TIME

A 16-foot-diameter wind turbine might produce one kilowatt of power in a steady 15-mph wind, depending on the design. One kilowatt of power produced continuously for 30 days amounts to 720 kilowatt-hours. This amount

---

*H. Wegley, M. Orgill, and R. Drake, *A Siting Handbook for Small Wind Energy Conversion Systems* (Battelle Pacific Northwest Laboratories, May 1978), 132 pp. Available from the National Technical Information Service, 5285 Port Royal Road, Springfield, VA 22161, order no. PNL-2521, $7.

of energy is about that used in the typical American home. The wind, however, is not constant. It is even more erratic than the average person would expect. The "speed-cubed" effect described in the previous chapter magnifies the effect of the fluctuating wind. For instance, a 20-mph wind has 2.37 times more power available than a 15-mph wind:

$$\frac{20 \times 20 \times 20}{15 \times 15 \times 15} = 2.37.$$

A 10-mph wind only had 0.30 times as much power as the wind blowing at 15 mph:

$$\frac{10 \times 10 \times 10}{15 \times 15 \times 15} = 0.30.$$

What if, instead of a 15-mph wind blowing continuously, half the time the wind blew at 20 mph and the other half of the time, at 10 mph? The average wind speed would be 15 mph, and the average wind power would be half of the two above numbers added together:

$$\frac{2.37}{2} + \frac{.30}{2} = 1.33.$$

So, in this simple example we find that 33 percent more power is available when the wind speed varies than when the wind is steady and has the same average wind speed.

Wind speed increases considerably with height. This is discussed in detail later in the chapter. To allow comparison of wind data, many of the recorded wind measurements in the world are taken at the standard height of 10 meters (32.8 ft).

Figure 3–3 shows average wind *speed* for each month at three locations in the United States.* Each monthly value is an average of records from many years. The resulting average monthly wind *powers* are shown in Figure 3–4. These are determined by averaging the cube of each hourly wind speed reading (i.e. 15 × 15 × 15) for each month. By comparing with Figure 3–3, the speed-cubed effect is, once again, very striking. For example, Big Springs, Nebraska, usually has its windiest month in April, with an average wind speed of 15.8 mph, and its least windy months in July and August, when the average wind speed is 12.0 mph. The average wind powers for these two cases are 451 watts per square meter† for April, and 210 for July and August.

The wind turbine owner who lives in an area with a high annual wind speed certainly has a great advantage. However, if his demand peaks in the season when the wind is at its minimum, the power requirement for a satisfactory

---

*We use data from several examples in this chapter to show how to use similar data for understanding your own site and estimating your wind energy potential.

†This refers to watts per square meter of rotor swept area of the wind machine. Refer to Chapter 5 for methods of calculating rotor swept area.

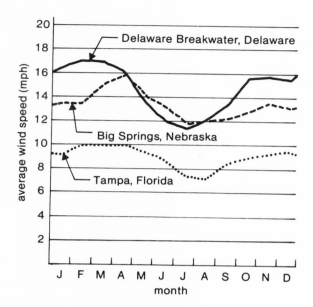

Figure 3-3: Average monthly wind speeds.

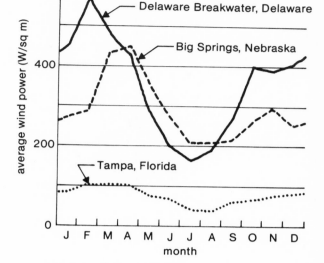

Figure 3-4: Average monthly wind powers.

wind system could be considerably higher than it would be if the annual average wind speed were less, but the seasonal wind speeds followed his demand. How to determine your power and energy needs is described in Chapter 4.

Storage battery costs are high enough that no one attempts to even out wind energy cycles from month to month with batteries. They are useful for making up hourly, daily, or weekly differences between supply and demand. Deficits over a month's time would have to be made up by alternative energy sources, such as an engine-driven generator, a wood stove, solar heating panels, or a connection to a utility power line.

Wind and Energy Roses

A *wind rose* plot is shown in Figure 3–5 for a weather station along the east side of Lake Michigan for one year. The length of each bar in these diagrams shows the percentage of time that the wind blows *from* that direction (toward the center of the circle). Each circle, or circular arc, represents 5 percent of the total time. The number at the end of each bar is the average wind speed for that direction. As an example, 11 percent of the time the wind blew from the northwest at the average wind speed of 18.5 mph (6.7 m/s), and 15 percent of the time from the south at 15.8 mph. The total length of all the bars adds up to 100 percent.

An *energy rose* is obtained by separately averaging the cubes of all the wind speed readings from each of the 16 directions. Figure 3–6 is the energy rose for the same case as the previous wind rose. The length of each bar gives the percent of wind energy from each direction. Again the total length of the bars is 100 percent. Notice the differences between the wind and energy roses. While the wind blows from the northwest 11 percent of the time, it is responsible for 21 percent of the annual available wind energy.

The battery capacity (or size of any other type of energy storage device) in

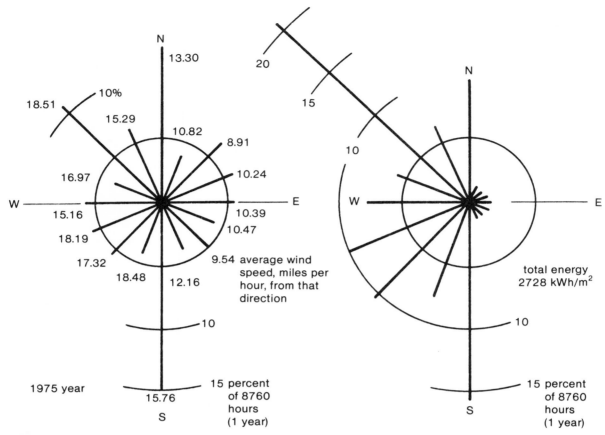

Figure 3-5: Wind rose for Muskegon Coast Guard Station, Michigan at 30 meters.

Figure 3-6: Energy rose for Muskegon Coast Guard Station, Michigan.

a wind energy conversion system will depend partly on the typical length of time the wind speed remains too low to generate an adequate amount of power. This waiting time for the wind to return is called the *return time*. The following table gives an example of four cities in Kansas, where a calm day is one where a speed of 10 mph is not reached. As an example, 10 times a year at Concordia, the return time is two days, and once a year, on an average, it is six days.

Location	Average Monthly Wind Power*	Total Calm (days/yr.)	Occurences of Return Times of:				
			2	3	4	5	6 days
Concordia	140	68	10	3	1	1	1
Topeka	157	52	8	2	1	0	0
Wichita	253	16	1	0	0	0	0
Dodge City	336	7	0	0	0	0	0

*Watts per square meter.

Notice that the number of consecutive calm days (wind speed does not reach 10 mph) increases rapidly as the average wind power decreases. The number of batteries required in a wind electric system will depend upon more weather factors than the return time. For instance, if the wind exceeds the cut-in

22

speed for only a fraction of a day and only partially recharges the batteries, the next return time must be very short or the batteries will be completely discharged.

Regular daily fluctuations in wind speed can be large or small. Figure 3–7 shows the typical daily fluctuations of hourly readings of wind speeds at three locations. Oak Ridge, Tennessee (5-year average record) is a southeastern U.S. interior location. Most of that area has a low average annual wind speed. The winds at the Muskegon Coast Guard Station, Michigan (one-year record) are considerably enhanced by a clear sweep from Lake Michigan and by strong lake breezes. The Livermore, California (one-month record) location is in a mountain pass into the great Sacramento–San Joaquin Valley, so a daily mountain-valley wind cycle occurs.

Everyone is aware of good and bad years for rain. Wind power also varies from year to year. Dodge City, Kansas has an average wind speed of 15.5 mph and an average wind power of 336 watts per square meter—about the highest for any city in the United States. In the 10-year period from 1955 to 1964, the yearly deviations from the average power were: +5, −9, +5, −22, +26, +3, −13, −15, +18.5, and +33.5 percent.

Figure 3-7: Sample daily wind variations.

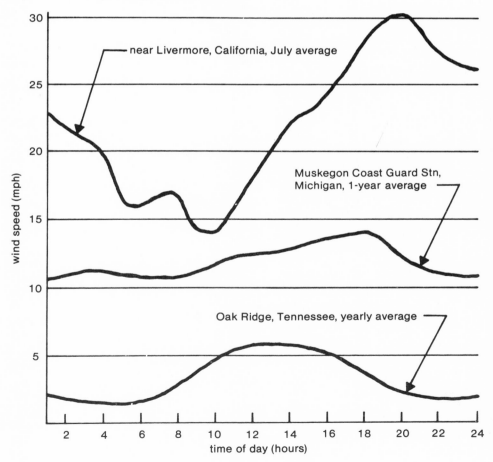

Determining Wind Power

How are monthly and annual wind powers determined from wind speed readings? The sophisticated way that has been used by meteorologists to produce power curves and energy roses has involved the use of computers to take the cube of regularly sampled wind speeds, usually hourly or every third hour, and then average the thousands of values to obtain these monthly or annual averages. Many of the hourly wind readings are available from the National Weather Service in a form ready to be processed by a computer.

There are several approximate methods that can be used to determine average wind power. First, the average speed can be used in the power equation given in Chapter 2. However, Figure 3–8 shows the large error that can occur if this is done. The shaded band in this figure represents the actual wind power divided by the power calculated from taking the cube of the average wind speed for 90 percent of the U.S. weather stations listed in Appendix 1. Like the example on page 20, the average of the cube of each of many fluctuating readings is considerably greater than the cube of the simple average wind speed. The effect is largest at low annual wind speeds. The band in this figure contains 90 percent of the values, so you have a 90 percent chance of being within this band. Taken from this figure, the average values are listed here:

Average annual wind speed, mph	8	10	12	14	16	18
Correction factor*	3.2	2.7	2.4	2.1	2.0	2.0

$$*\text{Correction factor} = \frac{\text{Average annual wind power}}{\text{Fictitious power calculated from average annual wind speed}}$$

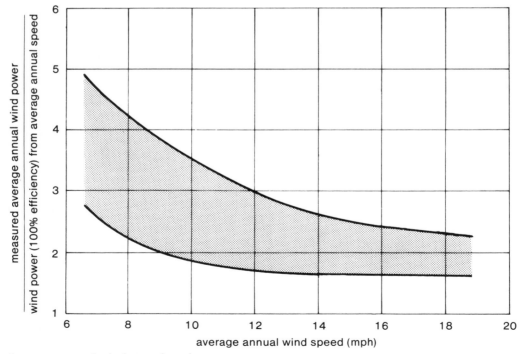

Figure 3-8: Average annual wind speed, mph.

A simple way to calculate the average annual energy for each square meter or square foot of rotor swept area is to use your average annual wind speed (the way to determine this is described in the last section of this chapter) in the power equation on page 17 along with the correction factor (page 24), times the number of hours (8760 in one year). Thus, the annual average energy generated, in kWh (kilowatt-hours), is approximately:

$$\text{annual kWh} = 8760 \times \frac{\text{watts}}{1000}$$

$$= \frac{8760}{1000} \times K \times e \times A \times V^3 \times \text{DRA} \times \text{DRT} \times (\text{correction factor})$$

For instance, if the two density ratio correction factors (DRT and DRA) are assumed to be 1.0, the average annual wind speed is 10 mph (so the correction factor from the table is 2.7), the windmill diameter is 15 feet (so the area $A = 3.14 \times 15^2/4 = 177$ square feet), and the average efficiency is 20 percent, then the approximate annual average energy generated is:

$$\text{annual kWh} = 0.876 \times 0.00508 \times 0.20 \times 177 \times 10^3 \times 2.7 = 4{,}253 \text{ kWh}$$

This is a simple way to obtain a value for your annual energy generation. It is, however, quite approximate, since the shaded band in Figure 3–8 indicates there is a 90 percent chance your value will be somewhere within 30 percent of this calculated value (above 16 mph the uncertainty drops from 30 to 20 percent). The data in Appendix 1 can be used to reduce this uncertainty.

A better way to determine the expected wind power from a specific wind turbine being considered is described in Chapter 5. The method uses a wind duration plot. Some typical wind duration curves are shown in Figure 3–9.

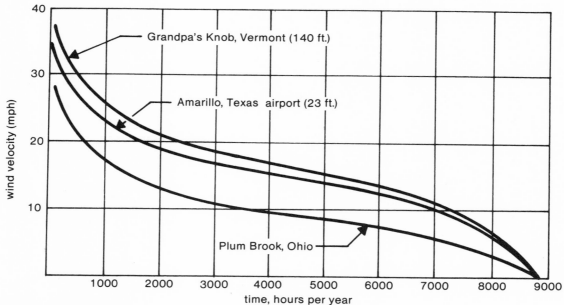

Figure 3-9: Wind duration curves.

Each point on these curves shows the number of hours in the year which the speed *equals* or *exceeds* the hours indicated directly below. For instance, a 10 mph wind speed is exceeded 3500 hours a year at Plum Brook, Ohio (where a large DOE-NASA wind turbine is being tested) and is exceeded 6950 hours a year at Amarillo.

AVERAGE WIND POWER DISTRIBUTION IN THE UNITED STATES

At over 1000 locations in the United States, a daily log sheet is filled out with hourly weather observations of the one-minute average wind speed and direction. These records are sent to the National Climatic Center in Asheville, North Carolina, where these one-minute averages for every third hour are entered onto computer magnetic tape. Various monthly and yearly summaries are prepared, and all the original data are stored in archives. Each station receives summaries of its data and these are usually available for inspection (this is described in more detail later in this chapter).

The data from 750 wind-recording stations in the United States and southern areas of bordering Canadian provinces have been processed to determine monthly averages of available wind power. The results are included in this report in Appendix 1. The stations are arranged with the states in alphabetical order and by region within the state. These data have not been corrected for varying heights of the wind anemometer* (the instrument used to measure wind speed). Also, distortions in the wind pattern by natural terrain features, trees, and buildings affect most of these locations. Most stations are at civilian or military airports. Very often, the anemometer location has been changed at least once over the years, new buildings have been erected, or even a highway overpass has been added nearby. No particular set of data can be blindly accepted as unaffected by obstructions. As an example, the energy rose for Moffett Field, California, one of the stations listed, shows 45 percent of the average annual wind power coming from the north-northwest. However, directly upwind of the anemometer and not very far away, stands one of the world's largest dirigible hangars! The wind and wind power measured from that direction would most certainly be different if measured upwind of the hangar.

Figure 3-10 shows very generally the wind power patterns for the continental United States. Typical open locations in the Pacific Northwest, southern Wyoming, Oklahoma-Texas Panhandle, and Cape Cod all average at least 200 watts per square meter at a height of 10 meters (32.8 feet). Two locations where data were previously presented, the east shore of Lake Michigan (Muskegon Coast Guard Station, Figures 3-5, 6, 7) and 50 miles east of San Francisco (Livermore, Figure 3-7) both have a high annual wind power—over 200 watts per square meter. These locations show here as areas of only medium and low wind power, respectively. This plot may be reasonably accurate for large open areas in the Great Plains but actually has little value in mountainous regions, as far as application to any specific site.

---

*Wind speed varies with height above ground. Anemometers mounted higher above ground will measure more wind.

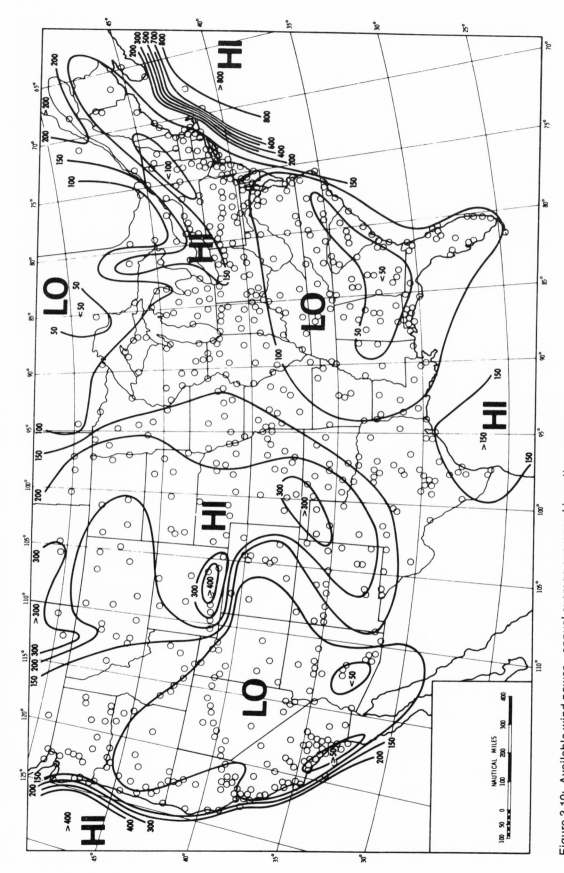

Figure 3-10: Available wind power—annual average expressed in watts per square meter. Courtesy of Jack Reed, Sandia.

27

Alaska and Hawaii are not shown, the latter having too few stations to allow this sort of plot to be developed. An example is made of Oahu later in this chapter. In Alaska, there are generally high wind areas along much of the coastline, but in general, the wind diminishes rapidly inland from the coast.

While Figure 3–10 gives a general impression of wind power available over the United States, the large table in Appendix 1 is much more useful.

WIND POWER VARIATIONS WITH HEIGHT AND LOCAL TERRAIN

From Figure 3–10 and the wind power records in Appendix 1, you can obtain a general idea whether your part of the country has good wind power. If you live in an area with apparently low wind power, certain types of hill terrain can double the local wind speed, and this will create an eightfold increase in available wind power. If, on the other hand, you are in a generally good wind area, your local winds may be disappointingly poor due to mountains, ridges, trees, or buildings. Generally, if you live in flat country with a meteorological station in the region, probably a fairly good estimate of your available wind power can be made before any wind survey is performed. However, if you live in hilly or mountainous country, or even in flat country with a considerable number of local obstructions, it is nearly impossible to estimate in advance the available wind energy. *Experience has shown that the typical person will greatly overestimate the local average annual wind speed.*

Effect of Height on Wind Power

We describe here how the wind power changes with height over reasonably flat country. The wind speed gradually increases with increasing height up to roughly 500 to 2000 feet above the earth's surface. Meteorologists call this region the atmospheric boundary layer. We will call how the wind changes with height the *wind speed profile*. Above it, the winds are more regular and only influenced by the largest geological features, such as mountains. Figure 3–11 shows typical winds over flat country at heights of 15, 30, and 100 feet near Sunnyvale, California. (For additional information on wind speed profiles, read Appendix 4.)

Figure 3–11 shows how irregular the winds are over a short time. It is not unusual for the instantaneous wind speed at a lower elevation to occasionally be greater than the wind speed at a higher elevation.

Figure 3–12 shows the average wind speed for each time of the day or night at 10 heights up to nearly a mile high above Oak Ridge, Tennessee. The data for all but the bottom two curves were obtained from balloon measurements. Five years of daily data were averaged to obtain these curves. While there are large changes in the average wind speed during the day at each height (except 160 meters), at any time during the daily cycle the *average* speeds increase with the height. By taking the average of each curve in the above figure and plotting that versus height, we have, in Figure 3–13, the average wind speed profile at Oak Ridge. This is an important type of curve, so let us

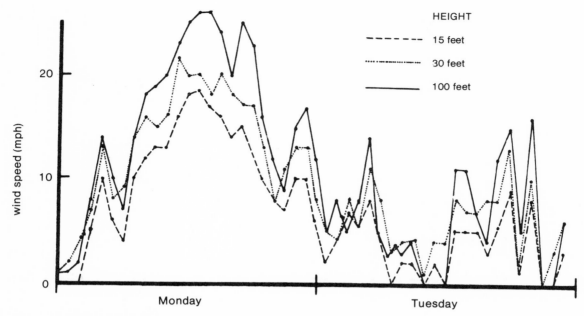

Figure 3-11: Hourly wind speeds at three heights.

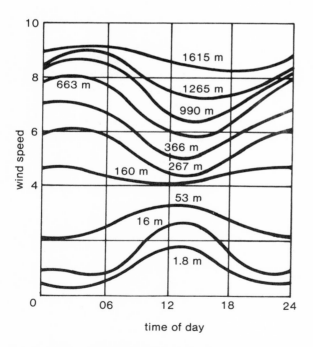

Figure 3-12: Daily curves of wind speed for several heights (expressed in meters) above the ground at Oak Ridge, Tennessee (five year averages).

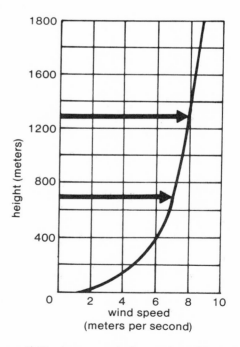

Figure 3-13: Average wind speed profile at Oak Ridge, Tennessee.

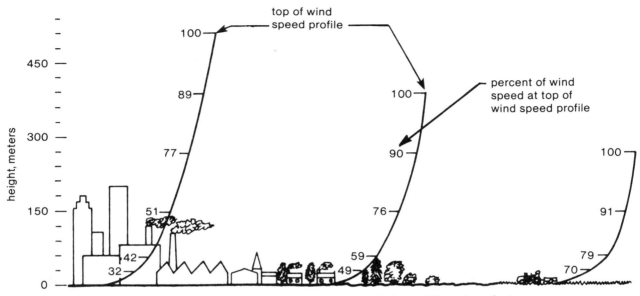

Figure 3-14: Atmospheric wind speed profiles change shape and height with surface features.

practice reading it. What are the wind speeds at the 600 and 1100 meter heights above the ground? Arrows have been drawn at these two heights. We move straight down from the tip of the arrows to the horizontal axis and read about 7 and 8 meters per second wind speed. Knowing the shape of your wind profile will help you select the best height for your wind turbine.

The location on top of the wind speed profile and the shape of the profile depends on (a) how flat the surface of the earth is, (b) the friction of the air trying to move across the surface of the earth, and (c) temperature differences along its path and up through the atmosphere. Figure 3–14 shows three wind speed profiles over flat terrain. Notice how the profile thickness increases with surface roughness. These profiles are really only correct when the wind is blowing strong enough to produce an appreciable amount of power from a wind turbine.

Effect of Regional and Local Terrain Features on Wind Power

Before delving into local terrain effects upon the winds, we should distinguish between two kinds of wind. The large weather features which cover the country at any time contain large scale patterns of winds. We experience this wind from above, or the *wind aloft*, after it is considerably diminished by the wind speed profile. *Local winds* are created from the ground up, so to speak, any time adjacent surfaces warm up at different rates and the winds aloft are not overpowering the local effects. Thus a sea breeze is created since land heats up much faster than the sea. A hillside receiving the morning sun warms up rapidly, while the valley below receives little direct sun and remains cold. This creates a wind. These local winds are in effect created by nature to reduce local air temperature differences; they do not have any appreciable effect on the winds aloft.

Local winds can only be described in general terms. If you live in an area with significant local winds, your awareness of these special situations will help you to capitalize on their wind enhancement effects.

Sea Breezes. During periods of light winds aloft in spring and summer, surface winds blow from ocean to land during the day (sea breeze) and in the reverse direction at night (land breeze). These breezes, as anyone near a large body of water knows, can be substantial winds. In winter, the land breeze may occur in the daytime as well as at night. Figure 3-2 roughly illustrates sea and land breezes. During a 24-hour period, the cycles of wind and temperature are much smaller over water than over land.

The development of a sea breeze is roughly as follows: Assuming there is no wind aloft, the sky is clear, the daytime has arrived, the sun will start heating the water and land. This heat is absorbed into several feet of water, but only into a fraction of an inch of earth, so the latter warms up much faster. The land heats the air at ground level, but this air gradually rises many hundreds of feet. The warmer air is lighter than the air over the sea so, as in the case of an open refrigerator door, the cold air rushes from the sea onto the land while the partially heated air far above the land moves out to sea to replace the incoming cold air. The circulation pattern has been completed; a sea breeze has been created. Starting as a local disturbance, this circulation pattern will extend many miles landward and seaward during each day.

A feeling for possible wind speeds in a sea breeze can be obtained from Figure 3-15. This shows the wind speed at 5 P.M. with no winds aloft in an area extending about 1 mile high by 20 miles out to the sea and 25 miles inland. In this example, winds from shoreline to about 11 miles inland are about 10 mph. Wind speeds of 15 mph caused by these sea breezes are common, even when winds aloft are still. Thus, a wind turbine at a seaside location can gain a great amount of power from this sealand breeze phenomenon and not rely solely on high winds aloft to be transmitted to the ground. Large lakes also create similar but smaller breezes.

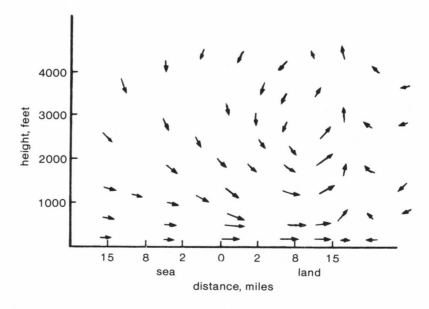

Figure 3-15: Example sea breeze winds, late afternoon, with no winds aloft.

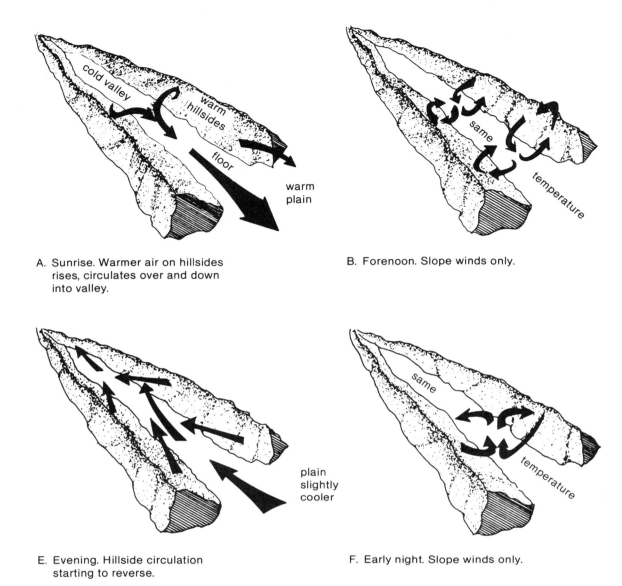

A. Sunrise. Warmer air on hillsides rises, circulates over and down into valley.

B. Forenoon. Slope winds only.

E. Evening. Hillside circulation starting to reverse.

F. Early night. Slope winds only.

Figure 3-16: Daily wind cycle in a valley facing a plain. No wind aloft.

Valley Winds. The complex nature of winds in valleys is briefly described here. When a strong wind aloft is blowing in a direction more or less parallel to a valley, there is a funneling effect. Winds are often stronger in the valley than over level country at these times, particularly when the valley narrows or its sides steepen. When the wind blows perpendicular to the valley, very complex flow patterns develop and often large areas in the valley will experience a great commotion in the air called *turbulence.*

Next, we consider what happens if the wind aloft is light. At night the air on the sloping sides of the valley will cool near the ground and, being heavier, will flow to the valley floor. When the slopes of the valley are warmed during the day, the wind will reverse direction. Complex combinations of these flows will occur as shown in Figure 3–16. The above effects cause most of the wind energy available in a valley to be aligned with the direction of the valley. Therefore, when siting a wind turbine, care should be taken to obtain the best location for capturing these winds.

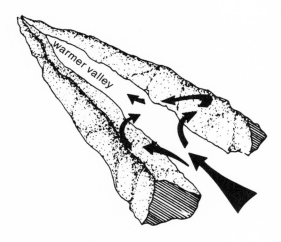

C. Early Afternoon. Breeze into valley and up hillsides.

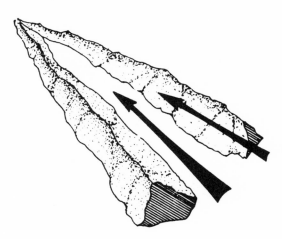

D. Late afternoon. Valley wind.

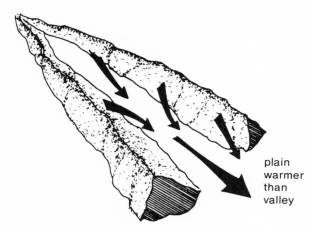

plain warmer than valley

G. Middle of night. Wind down hillsides, out onto plain.

H. Late night. Mountain wind.

If a valley narrows at its lower end, the cold air may drain out of the upper, broader end of the valley. A study of night wind profiles in a number of Vermont valleys indicated that maximum winds on most nights were found at heights of 100 to 1000 feet above the ground, often about two-thirds the height of the surrounding hills. The intensity of the wind gradually increases with increasing distance from the head of the valley.

In spite of the frequent valley winds, if prevailing wind directions are roughly at right angles to the valley, chances are that there is more wind energy available on the plateaus above the valley. The valley wind patterns sketched in Figure 3-16 may not contain a significant amount of recoverable wind energy.

It is easy to see that wind records from a meteorological station at another part of a valley from your location (or at a different distance from a coastline) may give little indication of your own winds.

Mountain Winds. The flow over a long, isolated mountain ridge that faces

33

the wind (and tends to block it) is another interesting case. Near the ground (at wind turbine height) the wind speed decreases as the toe of the ridge is approached, then speeds up to greater than the average in the region of the ridge line. If the sides of the ridge are very steep the increase in wind speed at the top will not be as great as for a ridge of moderate steepness. Also, the wind on the back side may be very turbulent and unsuitable for wind turbines. Near the toe of the hill on the upwind side the wind power may be reduced to 50 percent, and near the top it can be double the average wind speed depending on the slope of the sides.

What happens to the wind near the ends of a ridge that faces the prevailing winds? The Hawaiian island of Oahu provides an excellent example of this. Recently, a large computer was used to predict the flow over and around this island. For much of the year, such a strong wind blows across the island that the sea breeze influence is not particularly significant. This wind is confronted by a range of mountains 30 miles long (Figure 3-17) that can be described as a ridge about 2500 feet high, with occasional peaks several hundred feet higher. The ridge line is oriented only about 13° counterclockwise from a right angle to the prevailing wind direction. At each end, the ridge slopes down to sea level within a space of 5 to 10 miles.

The wind path lines at 500 feet above the surface of Oahu and the sur-

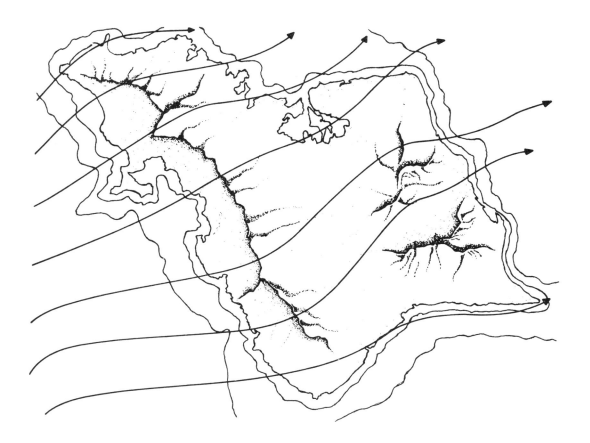

Figure 3-17: How the mountains of Oahu affect the wind.

rounding ocean are indicated in Figure 3-17. This ridge is very long—about 60 times as long as it is high—but even so, the wind path lines indicate that about one-third of the air approaching the island at the 500-foot height is deflected around the ends of the ridge. There are some good wind turbine sites along the top of the ridge line where the wind speed is about twice that of the approaching wind. The north and south ends of the ridge are particularly attractive sites, however, as the airflow that is deflected around the ends of the ridge speeds up to approximately twice the approaching wind speed for large regions at each end of the ridge. Thus, if your location is near the end of a ridge that tends to block prevailing winds, you may have considerably more wind power than your neighbors out on the flat land.

The wind near the top of an isolated hill is quite different than the wind over a long ridge. There is a strong tendency for the lower layers of the wind speed profile to split and go around the hill, particularly during the night and early morning when the ground has cooled off. For a wind turbine near the top, the hill acts quite a bit like a giant tower.

There is another characteristic of the wind that is helpful for increasing the wind power available on high hills. Reexamining Figure 3-12, the average daily wind curves above Oak Ridge, Tennessee, notice how the wind above 200 meters is greater at night than during the day while the reverse is true below 100 meters. This is a very general condition around much of the country. A wind turbine on a hilltop may produce power all night while below in the flat country the air will be nearly calm.

Structures and trees are likely to be in the vicinity of a wind turbine. How close can a wind turbine be placed to these, and how much penalty is paid if these obstacles cannot be avoided? Both loss of wind speed and the wind turbulence downwind of these obstructions are important. Wind turbine generators spin at high speeds and tend to have long, thin blades. This makes them much more susceptible to damage from wind turbulence than water-pumping windmills. Older water-pumping windmills are often found very close to trees and apparently are able to withstand the resulting turbulence. A rule of thumb generally used for wind turbine generator placement is: *If the location is not well above all surrounding obstacles, the wind turbine should be placed at least 10 obstacle heights or widths away from the obstacles.* This is a reasonable rule.

YOUR WIND POWER

You are indeed fortunate if you live in flat, open terrain and happen to have a neighbor with a wind turbine. You are doubly fortunate if he happens to be performing the same tasks with his machine that you wish to perform, such as pumping water or producing electricity. In effect, he has been measuring his wind power for a long time. The questions that you should consider are, how adequate is the wind for him, how does your demand compare with his, and how likely are you to have an appreciably lower (or higher) wind power per square foot (or square meter) than he has?

More likely, there is no wind turbine close enough to provide any useful experience for you. You will have to make a decision about how much time, effort, and money it is appropriate to invest to increase your knowledge of your wind. There are three parts to this process:

1. Making a preliminary estimate of your wind power.
2. Measuring your winds or wind power.
3. Comparing your measurements with nearby meteorological stations to determine your long-term average from your short-term data.

Before describing each of these, we will make a few observations. Step 1 is the very least you must do. That rough estimate of your wind power can be combined with results from a preliminary load survey (Chapter 4), equipment selection (Chapter 5) and a preliminary cost analysis (Chapter 6). You may find that you probably have only a fraction of the necessary wind available to make your investment a sound one and that only a minimal effort should be made to determine whether your wind power is much greater than your initial estimate.

Making a Preliminary Estimate of Your Wind Power

As a first step in making an estimate of your wind power, you should check for tall, unavoidable obstructions, particularly trees. Typical tower heights for wind generators in the 1- to 10-kilowatt class (the range you are most likely going to be interested in) are up to 100 feet. Two simple methods for measuring tree height, if the top is visible, are illustrated in Figure 3–18. For the first method, find a time in the morning or afternoon when the tree shadow falls across flat ground. Set up a vertical stick of known height and compare the length of the shadow from the stick to that from the tree. The equation for determining the tree height is:

$$\text{Tree height } (H) = \text{tree shadow } (S) \times \frac{\text{stick height } (h)}{\text{stick shadow } (s)}$$

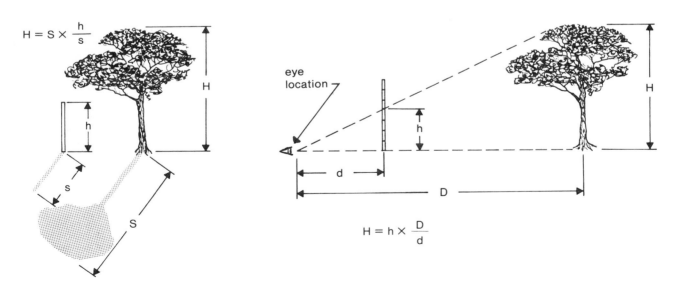

Figure 3-18: Two methods for estimating the height of a tree.

The second method for measuring tree height involves attaching a yardstick to a small pole or to the edge of some other fixed object so that you can simultaneously sight both the bottom and top of the tree without moving your head. The yardstick should be vertical. Note the distance along the yardstick between the lower and upper lines of sight; we call this distance h; the distance from your eye to the yardstick, distance d; and finally the distance from your eye to the tree, distance D. The equation for determining the height is then: $H = h \times D \div d$. This method can be used on rough terrain with changing slope, as long as the measurements are accurately made.

The second step in estimating your wind power involves finding the closest meteorological stations that are listed in Appendix 1. Compare their annual wind power averages. If they are at all close and there are no reasons (see previous discussion of terrain and obstruction effects on wind) why your value would not be about the same, you can take an average or place more weight on some of the readings than the others, according to your circumstances, to get a first estimate of your wind power. For open plains areas, this will work well, while in hilly or mountainous terrain, the results will most likely be poor. As an example, from Appendix 1, four meteorological stations in central Oklahoma (a typical plains area) within about 60 miles of a certain location have average annual wind powers of 263, 264, 174, and 167 watts per square meter. The average value is about 220, a very high value. On the other hand, an extreme case that can be taken from the tables in Appendix 1 and results in no useful data is the following: two meteorological stations are each 40 miles away from a fairly mountainous location. These two stations* have average annual wind powers of 420 and 45 watts per square meter!

The third step in obtaining a preliminary estimate of wind power is to simply start developing a better awareness of the wind and to make comparisons of the wind at your location with that where there are wind-measuring stations. The Beaufort Scale relates wind speeds to easily recognizable phenomena.

Hand-held wind anemometers, such as shown in Figure 3–19, are available at many boating, outdoors, and aircraft supply stores. These are the least expensive of all wind-measuring devices. One of these can be used to help calibrate your sense of the wind. Where around you do you find it to be especially windy or especially quiet?

Do you occasionally or even regularly pass by one of the listed stations in Appendix 1? Arrange to stop and look for the anemometer. Is it in a sheltered location or out in the open? Stay and eat your sandwich for lunch there. Get a feeling for their winds versus yours. At the very least, call them and ask about their anemometer location. Look for other anemometer stations that are not listed in Appendix 1, such as at small airports, forest fire and lookout stations, and colleges. Ask around.

Nearly all libraries will carry the government publication titled *Climatological Data*† for your state. It contains some wind data, and it lists all the local

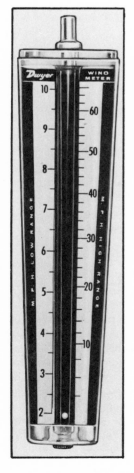

Figure 3-19: Hand-held wind anemometer.

---

*Point Arena and Santa Rosa, California, which are located in a mountainous, coastal region.

†Available from National Climatic Center, Federal Building, Asheville, NC 28801. Available on a subscription or single-copy basis. Subscription costs about $5 a year. Order by the state you want.

Beaufort Scale

	Observations	Speed (mph)
CALM	Calm. Smoke rises vertically.	0–1
LIGHT AIR	Direction of wind shown by smoke drift but not by wind vanes.	1–3
LIGHT BREEZE	Wind felt on face. Leaves rustle. Ordinary vane moved on wind.	4–7
GENTLE BREEZE	Leaves and small twigs in constant motion. Wind extends light flags.	8–12
MODERATE BREEZE	Raises dust and loose paper. Small branches are moved.	13–18
FRESH BREEZE	Small trees in leaf begin to sway. Crested wavelets form on inland waters.	18–24
STRONG BREEZE	Large branches in motion. Whistling in telegraph wires. Umbrellas used with difficulty.	25–31
NEAR GALE	Whole trees in motion. Inconvenience is felt when walking against the wind.	32–38
GALE	Breaks twigs off trees. Generally impedes progress.	39–46
STRONG GALE	Slight structural damage occurs (chimneys and roofs).	47–54
STORM	Seldom experienced inland. Trees uprooted. Considerable structural damage occurs.	59–63

weather observers for the National Weather Service. Most of these observers record only rainfall data, but they all receive this monthly publication.

The team that designed and built the Smith-Putnam wind turbine studied many promising wind sites in New England. They were mostly frustrated in their efforts to develop general rules for predicting the best wind sites. One powerful indicator that they did discover was that trees and plants can be greatly deformed in a consistent way by the wind. This is called *flagging*. They concluded:

1. Occasional very severe storms do not deform trees.
2. Tree deformation is a poor yardstick of maximum icing, although absence of breakage by ice may be significant.
3. Balsam trees are the best indicators of the mean wind speed in mountainous New England. Deformation begins when the mean velocity at a tree height of 30 to 50 feet reaches 17 mph. The other end of the scale is reached when balsam is forced to grow like a carpet, at a height of one foot, which indicates a mean wind speed of about 27 mph.
4. In this range between 17 and 27 mph, there are five easily recognized types of progressive deformation: brushing, flagging, throwing, clipping, and carpet.
5. Tree deformation is a sensitive indicator of the unpredictable wind flows through and over mountains. Local transitions from prevailing very high winds (which hold balsams to a carpet), to prevailing winds so moderate that the balsams reach normal growth without deformation, occur within a matter of yards of one another! Many a gardener knows that even moderate winds can have a strong effect on most vegetable plants.

Finally, combine all your information to make the best estimate of your wind power. Use this value for the first cut at sizing your wind system and estimating its costs.

Wind and Wind Power Measurement Equipment

To really measure your wind, you must mount a sensor on a pole high enough and far enough from buildings and trees to have a clear sweep of the wind. Ten feet is an absolute minimum height. Considering the time and money that will be invested in the survey, 25 to 50 feet is probably easily justifiable. Television antenna towers provide a good method to obtain these heights and can be as inexpensive as $1.50 to $2.00 per foot.

Most standard meteorological equipment for commercial use will appear to be quite expensive to the individual homeowner. However, new, relatively inexpensive items are rapidly being developed by manufacturers involved in the wind turbine market. Many of these items can also be rented from the manufacturer or distributor. For discussion purposes, this equipment can be divided into these types: (1) sensors for actually measuring the wind velocity, (2) meters for indicating the speed and/or direction, and (3) recording equipment that processes the data in one of several different possible ways and records the results.

The wind-cup anemometer (Figure 3–20) is the most popular type of sensor for surveys. It measures wind velocity but not direction. To measure the direction, a wind vane is required. Both pieces of data can be obtained with a third type, which has a propeller-type device attached to a tail vane. If you expect that the direction of the wind will be important to the placement of your wind turbine, then more than the wind-cup anemometer is advisable.

A note of caution must be made here. It is not unusual for an anemometer

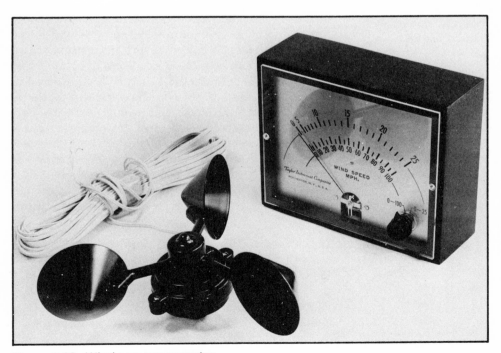

Figure 3-20: Wind cup anemometer.

to give readings that are considerably in error, particularly after extended use. A 10 percent wind speed error will produce about a 30 percent error in expected wind power, so some care is appropriate in selecting and maintaining the anemometer.

The simplest way of displaying the wind speed and direction is on meters. You must, however, read and log the values regularly, preferably hourly during your hours awake. These sensors can generate either a direct current signal, pulses, or an alternating current signal. If a dc signal is used by the sensor, the wire length will probably be important, while the pulse type is unaffected by wire length as long as the signal received has adequate strength.

The simplest recorder is a counter that displays the run-of-the-wind. It totals up the miles of wind that blow past the anemometer. Thus, if it records 240 miles in a 24-hour period, the average wind velocity is 10 mph. You only need to read it once a day, but it would be better to read it more frequently to establish daily wind patterns. A homemade wind vane can be used to advantage for estimating the wind direction.

Long-term recording devices need no attention on a daily basis. One type prints the wind speed and/or direction on a paper chart. Such continuous recordings as obtained at meteorological stations require expensive apparatus and are not necessary here. At a fraction of the cost is a small, adequate recorder that prints a dot onto a slowly advancing paper chart recorder indicating velocity every few seconds. A roll of paper will usually last a month or more. To use the data, one can eyeball an average for each time period, such as quarter-, half-, or one-hour intervals, and then record and process the data by hand. Owner's manuals and equipment manufacturer's publications detail simple methods for data reduction and use.

Several recorders are available that divide wind speeds into ranges (i.e. 0–3, 4–7, 8–11, etc.) and have counter displays for each range. For each time interval, such as one minute, one count is added to the counter for the average speed experienced during that period. The total counts in one month, for example, can be plotted as a wind duration curve (see Figure 3–9).

At least one manufacturer of WECS equipment presently sells a device that collects data essentially the same way as the previous recorders but then calculates the power that would result if some specific wind turbine were located there. There is just one display: total wind energy generated! The manufacturer will help you select a suitable wind turbine and use its characteristics in this recording device to perform the calculation. It can predict results for other windmills with a similarly rated wind speed. While this device is simple and direct, it is difficult to make correlations between its results and data from nearby meteorological stations.

Determining Your Wind Power

If you asked, "How long do I need to take wind data?" a meteorologist might tell you, "If you really want to know your wind, about five years is required!" He would be right and wrong. His answer could be right for some utility company which is considering the investment of millions of dollars in a group of wind-turbines and there is no long-term wind data for that region. His answer would not be appropriate if you are considering installing a $5,000 wind system.

Better questions are: "How much will I gain by taking data an extra week, an extra month, or an extra six months?" And "How much will I gain by using wind survey equipment with more capability?" Measuring your wind for one month and claiming you have determined your average wind power is as absurd as measuring your rainfall for one month, multiplying by 12, and claiming you now know your average annual rainfall. However, by comparison with local weather station records, a few months worth of rainfall records can often provide you with a good indication of your annual average rainfall. Likewise, your average annual wind power is best estimated by comparing your data with weather station records.

You might ask; "Where is the trade-off between time and money invested in a wind survey and the time and money invested in the windmill? What is the value of a wind survey to me? How much should I be willing to pay?" Let us look at three examples to get a feeling for what is involved.

Three Examples. For the first example, we assumed you live in flat country, about 20 or 30 miles from the nearest meteorological station. You have done a preliminary wind survey and you think you have a good site. The wind power at the nearest meteorological station is within ±40 percent of other stations within 100 miles. You feel that your wind power is probably within ±50 percent of that at this local station. We will assume that you have looked ahead at the next chapter and have evaluated your energy needs. You have tentatively selected several wind systems, the smallest one if your wind power is 50 percent greater than your estimate, and the largest windmill if your wind power is 50 percent less than the estimate. The largest unit has $(100 + 50) \div (100 - 50) = 3$ times the capacity of the smallest unit. They range in price from $3000 to $7000.

From your study of the economics involved (Chapter 6), you have decided that $7000 is your break-even point compared to extending the power lines and using public utility power. You feel that spending $700 for a survey would be well worthwhile. Say that the survey shows you have 25 percent more power available than your original estimate with a ±15 percent uncertainty. You would simply buy a $4000 system and be done with it. The $700 survey cost has in effect bought you assurance (in a sense, *insurance*) that you need not spend $7000. You have saved $2300 [7000 − (4000 + 700)].

As another example, suppose you have a summer cabin off in the woods and you use it approximately one month in a year. You are tired of hauling bottled gas and know that your tired, noisy, smelly, gummed-up engine-driven generator needs replacing. You have estimated the available power to ±40 percent. Since your entire need for electrical energy is only for recreation, when you hit an unusually calm spell of weather you could simply not go to the cabin. Using the high (140 percent) and low (60 percent) wind power and your estimated need, you come up with two systems, mostly composed of used equipment, with price tags of $1200 and $1600. The summer season is coming up and you are not there to take a survey and would have to contract it out. You decide to simply go ahead with a larger system since you feel you cannot get enough useful wind survey information for the few hundred dollars difference and the time allowed.

As a final example, consider a farmer or rancher who needs a lot of power. He has made a preliminary survey and selected two systems based on his estimate of maximum and minimum wind power that will cost $10,000 and

$25,000. Obviously, spending several thousands of dollars for a good wind survey will be a worthwhile investment.

The important features of all three of these examples are: (1) a preliminary estimate of wind power with an estimate of your accuracy; (2) an estimate of energy requirements (at least a preliminary estimate); (3) an estimate of your wind energy system low and high wind power cost; and (4) the maximum sum to place on the site survey. These four common features of the above examples are the first four steps shown in Figure 3–21, a diagram showing the steps to be taken in accomplishing a wind power survey.

Collecting and Using Data. Concerning the decision on the value of the wind survey, we return to the questions posed at the beginning of this section. "How much will I gain by taking data an extra week, month, or six months, and how much will I gain by using wind survey equipment with more capability?" These questions probably don't have any definite answers, but if you are interested in a small system, you live in flat country near a weather station with wind records, and your preliminary survey shows your

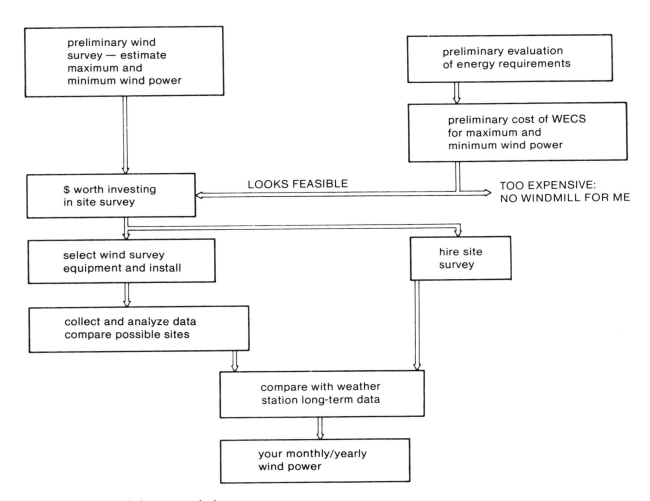

Figure 3-21: Determining your wind power.

estimated average annual wind speed is quite high, two or three months of data with one wind-cup anemometer will probably produce adequate results. For the potentially larger, more expensive system, more sophisticated wind survey equipment and longer data collection times are appropriate. Also, where the estimated average annual wind speed is less, there is a greater need for more than just run-of-the-wind readings. This is due to the larger scatter in wind power at the lower annual wind speeds shown in Figure 3–8, and described earlier in this chapter.

Equipment alternatives include renting or purchasing of the wind sensor equipment. Some wind turbine manufacturers/distributors will rent anemometer equipment. If you buy equipment, you can expect to recover some of your costs by selling it when you have finished your site survey. Also, you may wish to have a professional meteorologist, meteorology firm, or a knowledgeable distributor install the wind measuring equipment and perform the wind survey for you.

The next step in the process of determining your wind power (see Figure 3–21) is to compare your wind recordings with the readings that have been obtained simultaneously at nearby meteorological stations. You will then be able to estimate your long-term average wind power by using their summaries of years of data.

At each of the stations listed in Appendix 1, an hourly record of weather conditions is written out on form WBAN 10A, one sheet for each day. This sheet includes hourly observations of wind direction and speed. These are made by estimating on the hour the average wind speed for a one-minute period by observing a dial or strip chart recording. Visit or write your nearby station(s) to obtain copies of the records for the days of interest to you for comparison with your data. The WBAN 10A sheets (and the strip chart recordings) are sent to the National Climatic Center, and copies can also be ordered from them.*

Much of the data since the mid-1940s has been prepared for and processed by computers to obtain various long-term wind averages. Each station has a set of average monthly and yearly wind speeds for their location. These sheets are titled *Percentage Frequency of Wind Direction and Speed*. A sample from an exceptionally windy site is shown in Figure 3–22. It shows the percentage of the hourly readings in each of 11 different speed ranges (in knots) and 16 directions. In the right column is the mean wind speed (same as average wind speed) for each wind direction (remember that is the direction the wind blows *from*). Along the bottom are summations for each column and the average wind speed for the entire period. This last value will be essentially the same as listed in that station in Appendix 1.† You will at least want a copy of the annual summary and the sheets for your critical months when your expected demand is greatest compared to the available wind power. All the information you need for making up a wind rose (Figure 3–5) or a wind duration curve (Figure 3–9) is on this form. How to use this information to produce the wind duration curve is presented in Chapter 5.

---

*Federal Building, Asheville, North Carolina 28801.

†Small changes in the average monthly and yearly wind speeds can be caused by including more recent years of wind records than used for the results in Appendix 1. Any large change should be investigated. Have they relocated the anemometer or built new buildings nearby?

DATA PROCESSING DIVISION
ETAC, USAF
ASHEVILLE, N.C. 28801

SURFACE WINDS

PERCENTAGE FREQUENCY OF WIND DIRECTION AND SPEED
(FROM HOURLY OBSERVATIONS)

STATION	STATION NAME	YEARS	MONTH
45702	AMCHITKA ISLAND ALEUTIAN IS	44–50	JAN

CLASS	HOURS (L.S.T.)
ALL WEATHER	ALL

CONDITION

SPEED (KNOTS) DIR.	1-3	4-6	7-10	11-16	17-21	22-27	28-33	34-40	41-47	48-55	≥56	%	MEAN WIND SPEED
N	.1	.3	1.3	1.1	1.0	.5	.2	.1	.0	.0	.1	4.8	15.8
NNE	.2	.1	.5	.7	.9	.8	.6	.1		.0	.0	3.8	19.8
NE	.2	.1	.6	.8	1.2	.8	.7	.1	.0	.1		4.5	19.4
ENE	.1	.1	.3	.8	1.3	1.5	1.2	1.0	.3	.2	.9	7.7	30.0
E	.2	.4	1.1	1.9	1.5	1.0	.9	1.0	1.0	.3	.2	9.5	23.7
ESE	.1	.1	.5	1.0	1.3	1.3	1.1	.8	.8	.7	.1	7.9	27.4
SE	.1	.3	.6	.7	1.1	1.6	1.3	.9	.4	.3	.1	7.6	25.1
SSE	.0	.2	.3	.2	.3	.4	.4	.7	.5	.6	.3	3.9	32.1
S	.3	.3	.6	.9	1.0	.7	.7	.8	.5	.4	.2	6.4	24.9
SSW	.1	.1	.5	.6	.4	.7	.7	.3	.0	.2	.2	3.7	24.9
SW	.1	.2	.6	1.2	.9	1.1	.9	.5	.3	.2	.2	6.3	23.9
WSW	.0	.2	.8	.9	1.1	1.1	.9	.6	.3	.2	.0	6.1	23.0
W	.2	.3	.7	.9	1.3	1.4	.9	.4	.1	.1		6.3	20.7
WNW	.2	.3	.5	1.1	1.0	1.1	.7	.6	.2	.1		6.0	21.3
NW	.3	.4	1.7	1.9	1.4	1.4	.7	.6	.2	.0		8.7	18.1
NNW	.0	.2	.8	1.7	1.5	1.1	.4	.2	.0	.1		6.0	18.6
VARBL													
CALM												.9	
	2.3	3.7	11.4	16.4	17.3	16.6	12.2	8.7	4.8	3.7	2.1	100.0	22.9

TOTAL NUMBER OF OBSERVATIONS 5146

Figure 3-22: Sample of average wind data from an extremely windy location.

The method for using the data you collect from a wind-cup anemometer registering just the run-of-the-wind is quite simple. Read your meter at the beginning of each month. Follow the manufacturer's procedure for determining the miles of wind passing the meter during the previous month. Divide this number by the hours that have elapsed (for a 30-day month, 24 × 30 = 720) and you have your average wind speed for that month. Now, add up all the hourly wind speeds recorded at your nearest meteorological station for the month and divide that by the same number of hours. You have the average wind speed for the meteorological station for the same period. For instance, if your reading at the end of the first month is 5760, divided by 720 gives 8.0 miles per hour. If the sum of the weather station values for the same month were 6261, then the average wind speed would be:

$$6261 \times 1.15 \div 720 = 10.0 \text{ mph}$$

The 1.15 factor converts knots to miles per hour (all weather station readings are in knots). Your wind speed was 80 percent of that measured by the weather station, and your wind power is $(.8 \times .8 \times .8 \times 100) = 51$ percent of that at the station. Now, look in Appendix 1 at the average *annual* wind power for that station. Let us say for example, that this value is 200 watts per square meter. Your expected average annual wind power based on your single reading for one month would be $0.51 \times 200 = 102$ watts per square meter. This might be a quite satisfactory estimate, or it may be a very poor value depending on the terrain factors. If subsequent monthly comparisons show large differences in your wind power compared to the weather station wind power, the single month value was, of course, not good. However, if the values for the next couple of months are close to the first value, the average wind power based on these three months could be quite good.

CHAPTER 4

Power and Energy Requirements

In Chapter 2, we described the difference between work, energy, and power. This chapter will deal with two of these quantities: how much power is needed, and how long this power is required. This is the same as asking how much energy is needed.

As we stated in Chapter 1, the entire process of selecting a suitable wind system involves determining what power is available from the wind at your site, knowing what you need in the way of energy and power, and then matching these to arrive at a wind system that will do the job. We will break this discussion of power and energy requirements into two separate parts. The first will discuss electric load estimation, and the second will discuss mechanical load estimation, particularly for water pumping.

Before the discussion of power requirements, let us keep in mind that while it is well to calculate how much power you need, some consideration must be given to losses, or inefficiency. That is, if you figure how much electric power you need to run a light bulb, it will take a little extra, over and above the amount needed by the light bulb alone. The wires running from the generator to the light will waste some power because of electrical resistance. Friction is another example of inefficiency. This waste ends up as a power loss in the form of heat. Estimation of losses is an important part of your calculation of your power and energy needs.

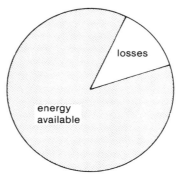

Figure 4-1: Some energy is always lost.

A pictorial view of energy and losses is shown in Figure 4–1. The smaller the pie slice representing loss, the more energy available to the user.

ELECTRIC LOAD ESTIMATION

Two different numbers will result from performing the simple calculations here—first, the power load, which is expressed as the number of watts, or kilo (thousand) watts, and second, the energy requirement, which is expressed in kilowatt-hours (kWh).

Electric power is a product of electric pressure, called volts, and current flow, called *amperes* (amps). Just as force times rate of motion equals power, usually expressed as horsepower, volts times amps equals power, expressed in watts. For example, a 12-volt battery that pushes 10 amps through a light uses electric power equal to $12 \times 10 = 120$ watts. Now suppose that the 120-watt light were left on for 10 hours. Then, the electric energy consumed would equal 120 watts times 10 hours, which would equal 1200 watt-hours, or 1.2 kilowatt-hours.

If a battery could store 2400 watt-hours of energy, then it would have a capacity to produce 120 watts of power for 20 hours. That is, 2400 watt-hours divided by 120 watts equals 20 hours. We will see more of these simple calculations as this chapter progresses.

To arrive at these types of numbers, it is necessary for you to determine two types of information:

- Which electrical devices you will use and how much power, in watts, they will draw.
- How long these devices will operate, say, in hours per month.

Later, we shall add a third item of information: at what time of the day the devices operate.

If it turns out that your wind system will provide power to electrical devices that you already use, and perhaps have used for some time, then load analysis becomes a simple task of checking all your electric bills for the last dozen or so months. It is a good idea to know how your electric bill changes with each month of the year to see what seasonal changes look like. In many cases, changes in the weather affect your energy use patterns. Here, you are not concerned with the dollar figure of the bill, but instead the actual demand figure expressed as kWh. Make sure that your utility company meter-reader has really read your meter, as occasionally utilities will estimate your use if the meter reader is behind schedule or is afraid of your BEWARE OF DOGS sign! Estimated kWh figures will not help you at all.

If you have not saved enough bills to check the demand, you can usually obtain a summary from your electric utility company. In either case, the utility bill will give you the monthly energy demand. It will not, however, give you the power demand in watts or kilowatts. For that, you will have to list the devices you use and determine the power requirement of each. Figure 4–2 will assist you. This figure will also serve to assist in estimation of energy demand if you do not have your utility bills.

Figure 4-2: Power and energy requirements of appliances and farm equipment.

Name	Watts	Hrs/Mo	kWhrs/mo	Name	Watts	Hrs/Mo	kWhrs/mo
IN THE HOME				Freezer, frost free	440	180	57*
				Fryer, cooker	1000–1500		5
Air conditioner, central			620*	Fryer, deep fat	1500	4	6
Air conditioner, window	1566	74	116*†	Frying pan	1196	12	15
Battery charger			1*	Furnace, electric control	10–30		10*
Blanket	190	80	15	Furnace, oil burner	100–300		25–40*
Blanket	50–200		15	Furnace, blower	500–700		25–100*†
Blender	350	3	1	Furnace, stoker	250–600		3–60*†
Bottle sterilizer	500		15				
Bottle warmer	500	6	3	Furnace, fan			32*†
				Garbage disposal equipment	¼–⅓ hp		½*
Broiler	1436	6	8.5	Griddle	450–1000		5
Clock	1–10		1.4*	Grill	650–1300		5
Clothes drier	4600	20	92*†	Hair drier	200–1200		½–6*
Clothes drier, electric heat	4856	18	86*†	Hair drier	400	5	2*
Clothes drier, gas heat	325	18	6*†	Heat lamp	125–250		2
Clothes washer			8.5*	Heater, aux.	1320	30	40
Clothes washer, automatic	250	12	3*				
Clothes washer, conventional	200	12	2*†	Heater, portable	660–2000		15–30
				Heating pad	25–150		1
Clothes washer, automatic	512	17.3	9*	Heating pad	65	10	1
Clothes washer, ringer	275	15	4*†	Heat lamp	250	10	3
Clippers	40–60		½	Hi Fi Stereo			9*
Coffee maker	800	15	12	Hot plate	500–1650		7–30
Coffer maker, twice a day			8	House heating	8000–15000		1000–2500
Coffee percolator	300–600		3–10	Humidifier	500		5–15*
Coffee pot	894	10	9				
Cooling, attic fan	1/6–¾ hp		60–90*†	Iron	1100	12	13
				Iron			12
Cooling, refrigeration	¾–1½ ton		200–500*	Iron, 16 hrs./month			13
Corn popper	460–650		1	Ironer	1500	12	18
Curling iron	10–20		½	Knife sharpener	125		¼*
Dehumidifier	300–500		50*	Lawnmower	1000	8	8*†
Dishwasher	1200	30	36*	Lighting	5–300		10–40
Dishwasher	1200	25	30*	Lights, 6 room house			
Disposal	375	2	1*	in winter			60
Disposal	445	6	3*				
				Light bulb, 75	75	120	9
Drill, electric, ¼"	250	2	5	Light bulb, 40	40	120	4.8
Electric baseboard heat	10000	160	1600	Mixer	125	6	1
Electrocuter, insect	5–250		1*	Mixer, food	50–200		1
Electronic oven	3000–7000		100*	Movie projector	300–1000		
Fan, attic	370	65	24*†	Oil burner	500	100	50*
Fan, kitchen	250	30	8*†	Oil burner			50*
Fan, 8"16"	35–210		4–10*†	Oil burner, 1/8 HP	250	64	16*
Food blender	200–300		½				
				Pasteurizer, ½ gal.	1500		10–40
Food warming tray	350	20	7	Polisher	350	6	2
Footwarmer	50–100		1	Post light, dusk to dawn			35
Floor polisher	200–400		1	Power tools			3
Freezer, food, 5–30 cu. ft.	300–800		30–125*	Projector	500	4	2*
Freezer, ice cream	50–300		½	Pump, water	450	44	20*†
Freezer	350	90	32*	Pump, well			20*†
Freezer, 15 cu. ft.	440	330	145*	Radio			8
Freezer, 14 cu. ft.			140*				

Symbol Explanation
 *AC power required
 ‡Normally AC, but convertible to DC

Notes: Lighting in this table is assumed to be incandescent—if flourescent, the wattage bulbs consume the same power but deliver 3 times as much light—flourescent bulbs also require AC, but can be converted to DC.

These figures can be cut by 50% with conservation of electricity.

SOURCE: *Energy Primer*, Portola Institute

Name	Watts	Hrs/Mo	kWhrs/mo
Radio, console	100–300		5–15*
Radio, table	40–100		5–10*
Range	8500–1600		100–150
Range, 4 person family			100
Record player	75–100		1–5
Record player, transistor	60	50	3*
Record player, tube	150	50	7.5*
Recorder, tape	100	10	1*
Refrigerator	200–300		25–30*
Refrigerator, conventional			83*
Refrigerator-freezer	200	150	30*
Refrigerator-freezer 14 cu. ft.	326	290	95*†
Refrigerator-freezer, frost free	360	500	180*
Roaster	1320	30	40
Rotisserie	1400	30	42*
Sauce pan	300–1400		2–10
Sewing machine	30–100		½–2
Sewing machine	100	10	1
Shaver	12		1/10
Skillet	1000–1350		5–20
Skilsaw	1000	6	6
Sunlamp	400	10	4
Sunlamp	279	5.4	1.5
Television	200–315		15–30
TV, BW	200	120	24*
TV, BW	237	110	25*
TV, color	350	120	42*
TV, color			100*
Toaster	1150	4	5
Typewriter	30	15	5*
Vacuum cleaner	600	10	6
Vacuum cleaner, 1 hr/wk			4
Vaporizer	200–500		2–5
Waffle iron	550–1300		1–2
Washing machine, 12 hrs/mo			9*
Washer, automatic	300–700		3–8*
Washer, conventional	100–400		2–4*
Water heater	4474	89	400
Water heater	1200–7000		200–300
Water pump (shallow)	½ hp		5–20*†
Water pump (deep)	⅓–1 hp		10–60*†

AT THE BARN

Name	Capacity HP or watts	Est. kWhr
Barn cleaner	2–5	120/yr.*
Clipping	fractional	1/10 per hr.
Corn, ear crushing	1–5 hp	5 per ton*
Corn, ear shelling	¼–2	1 per ton*†
Electric fence	7–10 watts	7 per mo.*†
Ensilage blowing	3–5	½ per ton
Feed grinding	1–7½	½–1½ per 100 lbs.*†
Feed mixing	½–1	1 per ton*†
Grain drying	1–7½	5–7 per ton*†
Grain elevating	¼–5	4 per 1000 bu*†
Hay curing	3–7½	60 per ton*
Hay hoisting	½–1	⅓ per ton*†
Milking, portable	¼–½	1½ per cow/mo.*†
Milking, pipeline	½–3	2½ per cow/mo.*†

Name	Capacity HP or watts	Est. kWhr
Sheep shearing	fractional	1½ per 100 sheep
Silo unloader	2–5 hp	4–8 per ton*
Silage conveyor	1–3 hp	1–4 per ton*
Stock tank heater	200–1500 watts	varies widely
Yard lights	100–500 watts	10 per mo.
Ventilation	1/6–⅓ hp	2–6 per day*‡ per 20 cows

IN THE MILKHOUSE

Name	Capacity HP or watts	Est. kWhr
Milk cooling	½–5 hp	1 per 100 lbs. milk*
Space heater	1000–3000	800 per year
Ventilating fan	fractional	10–25 per mo.*†
Water heater	1000–5000	1 per 4 gal

FOR POULTRY

Name	Capacity HP or watts	Est. kWhr
Automatic feeder	¼–½ hp	10–30 per mo*†
Brooder	200–1000 watts	½–1½ per chick per season
Burglar alarm	10–60 watts	2 per mo.*
Debeaker	200–500 watts	1 per 3 hrs.
Egg cleaning or washing	fractional hp	1 per 2000 eggs*†
Egg cooling	1/6–1 hp	1¼ per case*
Night lighting	40–60 watts	10 per mo. per 100 birds
Ventilating fan	50–300 watts	1–1½ per day*† per 1000 birds
Water warming	50–700 watts	varies widely

FOR HOGS

Name	Capacity HP or watts	Est. kWhr
Brooding	100–300 watts	35 per brooding period/litter
Ventilating fan	50–300 watts	¼–1½ per day*†
Water warming	50–1000 watts	30 per brooding period/litter

FARM SHOP

Name	Capacity HP or watts	Est. kWhr
Air compressor	¼–½ hp	1 per 3 hr.*
Arc welding	37½ amp	100 per year*
Battery charging	600–750 watts	2 per battery charge*
Concrete mixing	¼–2 hp	1 per cu. yd.*†
Drill press	1/6–1 hp	½ per hr.*†
Fan, 10"	35–55 watts	1 per 20 hr.*†
Grinding, energy wheel	¼–⅓ hp	1 per 3 hr.*†
Heater, portable	1000–3000 watts	10 per mo.
Heater, engine	100–300 watts	1 per 5 hr.
Lighting	50–250 watts	4 per mo.
Lathe, metal	¼–1 hp	1 per 3 hr.
Lathe, wood	¼–1 hp	1 per 3 hr.
Sawing, circular 8"10"	⅓–½ hp	½ per hr.
Sawing, jig	¼–⅓ hp	1 per 3 hr.
Soldering, iron	60–500 watts	1 per 5 hr.

MISCELLANEOUS

Name	Capacity HP or watts	Est. kWhr
Farm chore motors	½–5	1 per hp per hr.
Insect trap	25–40 watt	⅓ per night
Irrigating	1 hp up	1 per hp per hr.
Snow melting, sidewalk and steps, heating cable imbedded in concrete	25 watts per sq. ft.	2.5 per 100 sq. ft. per hr.
Soil heating, hotbed	400 watts	1 per day per season
Wood sawing	1–5 hp	2 per cord

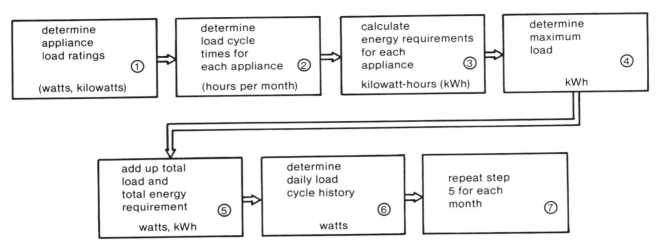

Figure 4-3: Estimating your energy requirements.

We will follow the blocks in Figure 4–3 to apply a logical sequence to the following discussion.

BLOCK 1. Determine the appliance load rating, expressed in watts or kilowatts.

Example: Brand C electric motor is a one-horsepower motor. Its electrical load when operating is 860 watts, and when starting, 1400 watts for one second. These data are found: on the appliance data plate, by writing to the manufacturer, by testing an appliance yourself, or from Figure 4–2 of this book. You can easily test the appliance if you presently have electric service. Watch your electric utility meter, which measures energy usage. Usually it contains a slowly spinning disc, and some number of revolutions of it indicates that one kWh has been consumed. Ask your power company what each revolution means. Turn off all other appliances so the meter stops. Then turn on the appliance you wish to rate and time the spinning disc.*

BLOCK 2. Determine the load cycle time and the number of hours the load will operate on a monthly basis.

Example: Brand D refrigerator will operate an average of 15 hours per month. Note that this information depends on how well insulated the refrigerator is, the number of times the door is opened, how much bulk will be stored, and the room temperature.

BLOCK 3. Determine the appliance's monthly energy requirement. This is calculated by multiplying together the data from blocks 1 and 2 above. The result is in watt-hours, or kilowatt-hours.

Example: A color television is determined in step 1 to require 350 watts. You know that it will be used for 150 hours per month, so:

350 watts times 150 hours equals 52,500 watt-hours.
To get kilowatt hours: 52,500 ÷ 1000 = 52.5 kilowatt-hours (kWh)

---

*This will not be adequate for obtaining the starting load, but that is not necessary for these load calculations.

BLOCK 4. Determine maximum load. This calculation will result in your maximum power demand, expressed as kilowatts (kW), which would occur if all of your appliances were operating at the same time.

Example:

Item	Power (watts)
TV	200
Coffee pot	894
Dishwasher	1200
Refrigerator	300
Water heater	6000
Maximum Load:	8594 watts = 8.59 kW

BLOCK 5. Determine your total monthly energy requirement. Total energy is expressed in kilowatt-hours (kWh). This is the energy you must supply each month or pay for when supplied by your utility company.

Example:

Item	Energy requirement (kWh)
TV	24
Coffee pot	9
Dishwasher	36
Refrigerator	30
Water heater	300
Total energy requirement	399 kWh per month

(Note: The above examples are selected at random from Figure 4-2 and do not necessarily represent a typical household load.)

We shall return to monthly energy requirements shortly, but first let us look more closely at a daily breakdown of your electrical load.

BLOCK 6. Determine your daily load cycle history. This calculation will give you a much better estimate of your actual electric load demand. Simply adding up all of the loads, as in the above example, assumes that all appliances will be on at the same time and gives a worst case figure, but does not reflect a real case. To arrive at a load cycle history, you must make estimates of the time of day your devices will be on and for how long.

This estimate may be as accurate or as rough as necessary. How accurate you decide to be in making estimates will depend entirely on your assessment of the importance of this calculation. For an accurate estimation, it will be necessary to actually monitor any items that operate on a cyclic basis, such as refrigerators. For less accuracy, it may be reasonable to assume such loads "on" continuously. As a first example (not representative, but illustrative of the thinking here), we shall separate items by their nature: those which you control, and those which operate automatically. For this example, items that operate automatically shall be assumed to operate *continuously*. The other loads will require estimation of operating cycles, as listed on page 52.

Automatic Items	*Load (watts)*	*Time*
Refrigerator	300	Continuous
Water heater	6000	Continuous
User-controlled Items		
TV	200	4 hours per day as follows: 1 hour: 8 A.M.–9 A.M. 3 hours: 6 P.M.–9 P.M.
Coffee pot	894	20 minutes per day 7:30–7:50 A.M.
Dishwasher	1200	1 hour per day: 5 P.M.–6 P.M.

From this table, we can see a *base load*, that is, a continuous load equal to 6300 watts, with peak loads going as much as 1200 watts higher. Again, this is not a representative example, but compare it with the following example.

If we made a simple graph of this load, it would look something like Figure 4–4.

For a second example, let us be more realistic. The assumption that all automatic items are continuous loads should be adjusted. Since the data were originally extracted from Figure 4–2, let us look at that chart again.

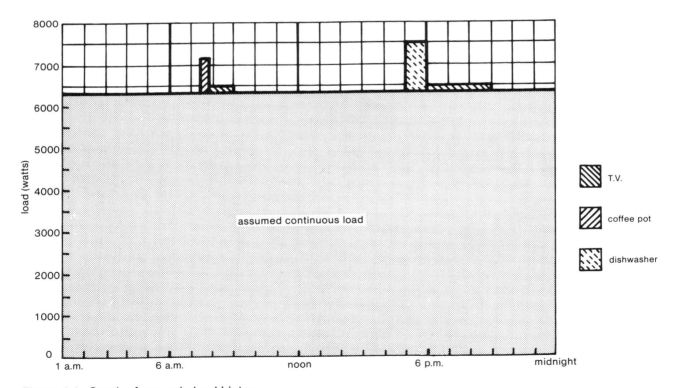

Figure 4-4: Graph of example load history.

Notice that the refrigerator is listed as 200–300 watts, for 25 to 30 kWh per month. Using 300 watts and 30 kWh (or 30,000 watt-hours) per month, we can calculate the hours per month this device operates:

30,000 watt-hours per month ÷ 300 watts = 100 hours per month.

Now, assuming a 30-day month, 100 hours per month ÷ 30 days = 3.3 hours per day. This is the estimated number of hours per day this refrigerator will operate. Now we must guess when, and for how long during each cycle it operates. A safe guess is that it cycles most during mealtimes.

For the water heater, a similar calculation should be made: 300,000 watt-hours per month ÷ 6000 watts ÷ 30 days = 1.6 hours per day.

Automatic Items	Load	Time
Refrigerator	300	3.3 hours per day: 1.1 hours each 7 A.M., noon, 5 P.M.
Water heater	6000	1.6 hours per day: 0.8 hours each 8 A.M., 6 P.M..

With use-controlled items similar to the first example, results from the new graph (Figure 4–5) will be somewhat closer to reality.

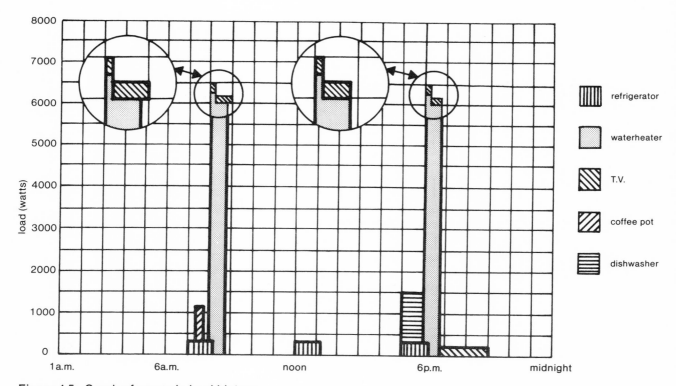

Figure 4-5: Graph of example load history.

Performing the type of analysis in Figure 4-5 may not actually be necessary for your energy-requirement estimates, but it is a good way to understand the nature and characteristics of the electric load you expect.

BLOCK 7. Determine your monthly energy requirements. A previous example illustrated how a monthly energy requirement of 399 kWh was established. This could have been any month. For some months, a heavy demand for heating may raise electrical consumption, while in others, air conditioning will prevail. Thus, you must complete a demand analysis for each month of the year. A graph plotted from your totals would look like Figure 4-6.

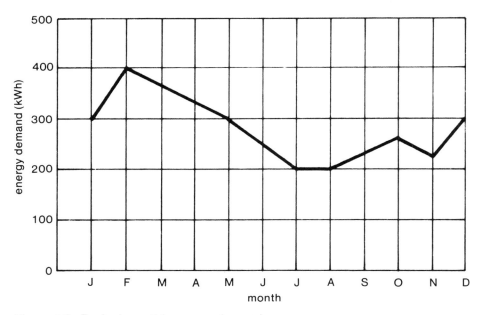

Figure 4-6: Typical monthly energy demand.

MECHANICAL LOAD ESTIMATION

Estimation of mechanical load may be as simple as reading the data plate on a device you expect to drive mechanically by wind power, and can be as complex as calculating the horsepower required to pump water through some pipes. This section will deal primarily with estimation of power required to pump water.

Figure 4-7 illustrates a complete water pump and storage system. In any system such as this one, you are expected to know the pump depth and tank height. Add these two together and you get water head—the maximum height to which the pump must lift water (we assume the tank is not pressurized).*

Water head = pump depth + storage tank height

---

*For a pressurized tank add 2.31 feet of water head for each psi.

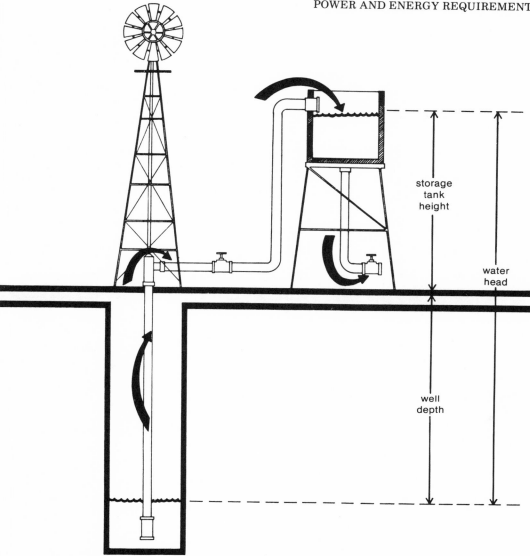

Figure 4-7: Water pump system diagram.

(Where water head is measured in feet, pump depth is measured in feet from ground surface [minimum allowable water height] and tank height is measured from ground surface to top of water outlet at tank.)

> *Example:* Pump depth   200 feet
> Tank height   70 feet
> Water head = 200 + 70 = 270 feet

Notice that while the water is being lifted to the total height of 270 feet as in the example, it must pass through pipes that are considerably longer, unless the tank is directly on top of the well. The loss of pressure (*head loss*) from water flowing through the pipes will increase the amount of load on the pump. We calculate the head loss using the *head loss factor* from the graph on Figure 4–8. This factor is the feet of head loss per hundred feet of pipe

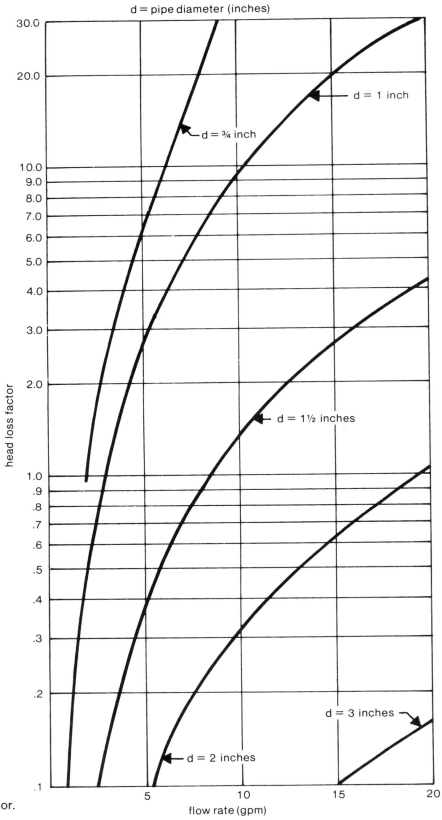

Figure 4-8: Head loss factor.

56

run. Thus, you need to measure total length of pipe run and know flow rate measured in gallons per minute and the pipe diameter.*

From Figure 4–8 a value of head loss factor can be determined that, when you have measured pipe run, can be converted to head loss.

Head loss = head loss factor × pipe run ÷ 100
(When pipe run measured in feet.)

Note: This value is for steel pipe and does not include valve and fitting losses. If several hundred feet of pipe length is involved, fitting losses can be neglected. To include them in an approximate way, count up the number of tees and elbows. For 1-inch pipe add 3 feet of pipe *length* for each fitting and for fully open valves, add none for a gate, 12 feet for an angle, and 30 feet for a globe valve (the usual spigot-type valve). This is then the total equivalent pipe length. For other pipe sizes, proportion these lengths to the pipe size; i.e., twice these values for 2-inch pipe. For smooth plastic pipe reduce the head loss factor by 40 percent.

> *Example:* Maximum pump capacity – 5 gpm
> Pipe run – 250 feet
> Water head – 250 feet
> Pipe diameter – ¾ inch
> Two 90° elbow fittings installed

From Figure 4–8, find 5 gpm flow rate on the horizontal (bottom) line. From this point, go straight up to the ¾-inch pipe curve line. From there, look straight to the left to read head loss factor = 6.0 on the vertical scale. This is 6 feet per 100 feet of pipe run. Then total head loss factor = 6 × 250 ÷ 100 = 15 feet. For the two elbow fittings add 2.3 feet each. Then head loss = 15 + 4.6 = 19.6 feet. From here, we calculate *total head*, which is the actual load presented to the pump.

Total head = water head + head loss
(Where all factors are measured in feet.)

Continuing the example, total head loss = 250 + 19.6 = 269.6 feet, or 270 feet.

Because of head loss, the pump is loaded as if it has to pump water 19.6 feet higher than it really does.

Now we can calculate horsepower required by the pump and supplied by the wind turbine. From Figure 4–9, you can read the theoretical horsepower (no losses) knowing the total head and flow rate. Continuing our example, for total head = 270 feet, and 5 gpm, the theoretical horsepower = 0.3. This is the horsepower supplied *by* the pump. Now you must calculate horsepower supplied *to* the pump by the wind turbine. For most well-maintained or new

---

*Figure 4–8 assumes standard ("schedule 40") steel pipe. For other diameters than those listed, the head loss at the same flow rate is proportional to the fifth power of the ratio of the pipe diameters. Pipe diameter is approximately the internal diameter. The outside diameter will be ¼ to ½ an inch larger.

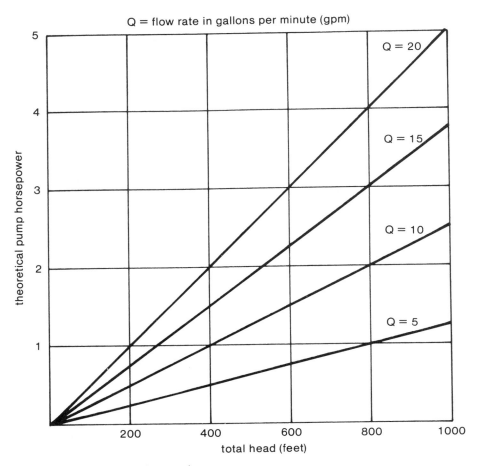

Figure 4-9: Theoretical pump horsepower.

piston pumps installed on wind turbines, assume a pump efficiency of 70 percent.

Then:

$$\text{Wind turbine horsepower} = \text{pump horsepower} \div 0.7$$

Example: Pump horsepower calculated in previous example = 0.3. Then wind turbine horsepower = 0.3 ÷ 0.7 = 0.43. Thus, a ½-horsepower wind turbine can pump water at a little more than 5 gpm up a total height of 250 feet, through 250 feet of ¾-inch iron pipe with two elbow fittings installed.

At this point in our calculations, we have developed a method to predict how much horsepower is needed. This is a power requirement, but, as with electrical systems, we need to know horsepower-hours (hp-hr), the energy requirement.

For this, you must estimate your daily (or monthly) water requirements in gallons just as you would estimate electrical requirements.

Use Figure 4–10 to add up the gallons of water you need daily (adjust these according to your experience). Multiply values by 30 for monthly calculations.

Effect of External Temperature on Water Consumption

Water Consumption of Hogs
(Pounds per Hog per Hour)

Temperature (°F.)	75–125 lb. hogs	275–380 lb. hogs	Pregnant Sows
50	0.2	0.5	0.95
60	0.25	0.5	0.85
70	0.30	0.65	0.80
80	0.30	0.85	0.95
90	0.35	0.85	0.90
100	0.60	0.85	0.80

Water Consumption of Dairy Cows
(Gallons per Day per Cow)

Temperature	Lactating Jerseys	Lactating Holsteins	Dry Holsteins
50	11.4	18.7	10.4
50–70	12.8	21.7	11.5
75–85	14.7	21.2	12.3
90–100	20.1	19.9	10.7

Water Consumption of Hens
(Milliter per Bird per Day)

Temperature	White Leghorn	Rhode Island Red
70	286	294
80	272	321
90	350	408
100	392	371
70	222	216
70	246	286

Water Consumption of Sheep
(Pounds of Water per Day)

On range or dry pasture	5–13
On range (salty feeds)	17
On rations of hay and grain or hay, roots and grains .	0.3–6
On good pasture	Little (if any)

Water Consumption of Pigs
(Pounds of Water per Day)

Conditions

Body Weight—30 lbs.	5–10
Body Weight—60–80 lbs	7
Body Weight—75–125 lbs	16
Body Weight—200–380 lbs	12–30
Pregnant Sows	30–38
Lactating Sows	40–50

Water Consumption of Chickens
(Gallons per 100 Birds per Day)

Conditions

1–3 weeks of age	0.4–2.0
3–6 weeks of age	1.4–3.0
6–10 weeks of age	3.0–4.0
9–13 weeks of age	4.0–5.0
Pullets	3.0–4.0
Nonlaying hens	5.0
Laying Hens (moderate temperatures)	5.0–7.5
Laying Hens (temperature 90°F)	9.0

Water Consumption of Growing Turkeys
(Gallons per 100 Birds per Week)

Conditions

1–3 weeks of age	8– 18
4–7 weeks of age	26– 59
9–13 weeks of age	62–100
15–19 weeks of age	117–118
21–26 weeks of age	95–105

Water Consumption of People

Average person: 75 gallons per day

Lawn 0–200 gallons per 1000 square feet, every other day.

Water Consumption of Cattle

Class of Cattle	Conditions	Pounds per Day
Holstein calves (liquid milk or dried milk and water supplied)	4 weeks of age	10–12
	8 weeks of age	13
	12 weeks of age	18–20
	16 weeks of age	25–28
	20 weeks of age	32–36
	26 weeks of age	33–48
Dairy Heifers	Pregnant	60–70
Steers	Maintenance ration	35
	Fattening ration	70
Range Cattle		35–70
Jersey Cows	Milk Production 5–30 lbs/day	60–102
Holstein Cows	Milk Production 20–50 lbs/day	65–182
	Milk Production 80 lbs/day	190
	Dry	90

SOURCE: Water, *Yearbook of Agriculture*, U.S. Department of Agriculture *1955*.

Figure 4-10: Water requirements.

Hypothetical example:

2 persons:	75 gal/day × 30 =	2250 gal/mo
1 beef cow:	12 gal/day × 30 =	360 gal/mo
20 chickens:		
(4 gal/day ÷ 100) ×	20 ÷ 100 × 30 =	24 gal/mo
lawn (1000 ft²):	160 × 15 =	2400 gal/mo
	TOTAL	5034 gal/mo

Using the water requirement data, you can calculate hp-hr.

Example: Assume the ½-horsepower pump of previous examples pumps at an average rate of 5 gpm (this rate varies with the changing wind speed).
Calculate hp-hr per month: 5034 gallons per month ÷ 5 gallons per minute = 1,007 min/mo, or 1,007 ÷ 60 = 16.8 hours per month required at an average flow rate of 5 gpm. Then energy required (monthly hp-hr) = 16.8 × ½ = 8.4 hp-hr.

ENERGY STORAGE: WATER SYSTEM

The water pump system of the previous example was required to produce 8.4 hp-hr per month at an average flow rate of 5 gpm. Since wind speed varies with the time, you can expect that, at certain times of the day (or month) wind will blow strong enough to pump water faster than 5 gpm, while at other times flow rate will be less or no water will flow at all. The user may wish to use water at times when no wind is blowing, while at other times the wind turbine tries to pump water that is not being used.

To make up the difference in various conditions, a storage tank is used. Water pumped up to that tank represents energy stored.

To calculate energy storage requirements for a water system such as illustrated in Figure 4–7, you need to know the maximum number of days for which you must store water. These data come from your wind site survey (Chapter 3), where data such as maximum number of windless days become available.

Example: From the previous example, let us total the water demands on a daily basis:

2 persons	75	gal/day
1 beef cow	12	gal/day
20 chickens	1	gal/day
lawn	160	gal every other day

TOTAL: Low daily (without watering the lawn) = 88
High daily = 248
Average daily = 168

Hypothetical site data: maximum = 6 windless days.
If this occurs in January, when no lawn water is required, use the low daily value:

88 × 6 = 528 gallons storage requirement

If the windless day (month) occurs in July, use the high value:

$$248 \times 6 = 1,488 \text{ gallons storage requirement}$$

Depending on the relative importance of lawn water and cost of water storage, you might be inclined to weigh the two against each other in the selection of a water tank size.

Figure 4–6 illustrates a month-by-month, electric-energy demand curve. Figure 4–11 is a slightly more complex presentation of another energy demand curve for a hypothetical wind turbine installation. On the same graph is plotted the energy supplied to the user by his wind turbine generator.

Figure 4-11: Comparing monthly energy supply and demand.

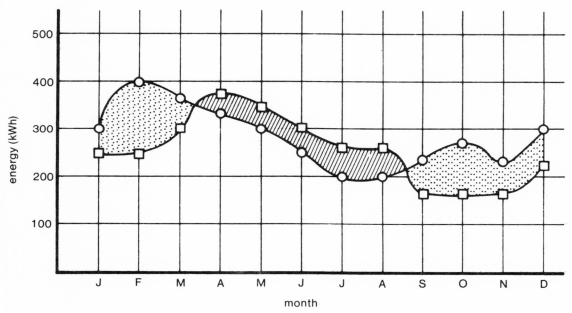

Month	Supply kWh	Demand kWh	Difference kWh
J	250	300	50 needed
F	250	400	150 needed
M	300	350	50 needed
A	330	300	30 extra
M	300	250	50 extra
J	280	200	80 extra
J	280	200	80 extra
A	250	200	50 extra
S	180	200	20 needed
O	180	250	70 needed
N	180	200	20 needed
D	210	250	40 needed
Totals	2,990	3,100	

O = demand
□ = supply
▒ = supply below demand
▨ = excess power available

Notice (Figure 4–11) that from January through March, demand is higher than supply. From April to September, supply is greater than demand, and then from September to January, demand is greater again. Demand, on an annual basis, is 110 kWh greater than supply. Either demand must be reduced by 110 kWh/year, or supply increased by that amount, which is only a (110 ÷ 2990) × 100 = 3.7 percent increase. Increasing supply can come from a larger wind turbine, a backup energy source, such as a gasoline-powered generator, or more wind. Using the wind profile information in Chapter 3, we could determine how much to increase the height of the wind turbine tower to obtain the needed additional wind (about five feet).

Would it be possible to store excess energy from April through August, and use this from September into the following February when it would be exhausted? We will assume each year is the same. We calculate the actual storage requirement by adding up all the monthly surpluses in Figure 4–11:

April	30 kWh
May	50
June	80
July	80
August	50
TOTAL	290 kWh

In Chapter 5, you will learn how many batteries would be required to hold this much energy. For now, it is sufficient to say that for a 290-kWh battery bank, using 6-volt golf cart batteries rated at 200 amp-hours each, we would need 242 batteries. At a cost of, say, $30 each, this energy storage system would cost $7260. Obviously, this amount would buy a much larger wind turbine and nearly eliminate the requirement for batteries. You may also choose to use an auxiliary generator during periods of insufficient wind to minimize the storage requirement. These options are described in Chapter 5.

As with the water storage example, it is important to know the maximum number of windless days. You also need to know that a suitable surplus of energy will be generated prior to these windless days, or you will start the windless cycle with dead batteries.

Example: Average daily energy usage – 5 kWh
Maximum number of windless days – 5
Energy storage requirement = 5 kWh per day × 5 days = 25 kWh

Again, using 6-volt, 200 amp-hour batteries, 21 batteries would be required at a cost of about $630, if the $30 per battery assumption is correct.

CHAPTER 5

The Components of a Wind Energy Conversion System

A wind-turbine dealer should be able to advise you on the components for your wind energy system. This chapter will provide you with the basics on each component you purchase. It will help you ask the right questions and better understand any WECS brochures you receive. The wind turbine is considered first. Why do some wind turbines only have two or three narrow blades while others have many wide blades? What are the basic choices in types of wind-turbine rotors? These items are discussed, and a method is presented for comparing wind turbines by calculating the wind energy each will generate. For wind electric system owners the two sections that follow on generators and energy storage devices will be useful. Finally, towers, inverters, backup equipment, and the typical efficiencies of the various components are described.

In Figure 5–1, the *windwheel* (or rotor) is the device that actually processes the wind and converts it to mechanical power. To visualize how these blocks and arrows translate to actual hardware look at Figure 5–2.

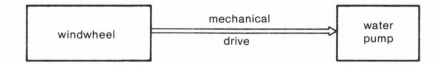

Figure 5-1: Simple wind system.

Here, the "windwheel" is a parachute that uses a drag force caused by the wind to tug on a rope, which drapes over a pulley and is connected to a bucket (pump). Let us analyze the advantages and disadvantages of this wind system.

The advantages are:

- Simple and easy to understand, low initial cost.
- Can be easily folded and stored during times of poor wind condition.
- Favorable starting torque for lifting water from the well.

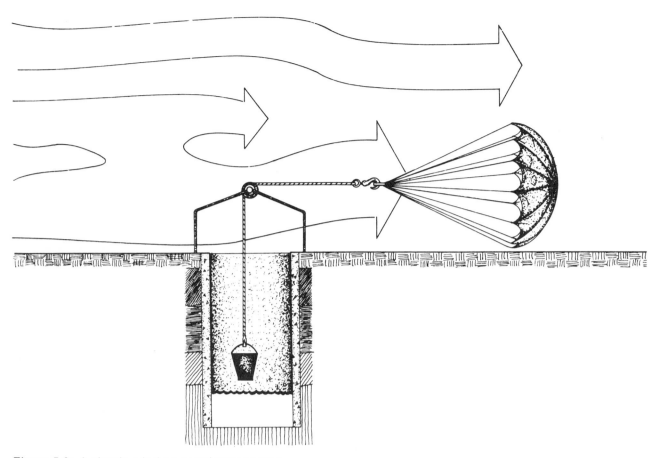

Figure 5-2: A simple wind-powered water pump.

- Easily repaired.
- Does not require tall towers.

The disadvantages are:

- Requires constant attention to operate.
- Operates on a cyclic basis requiring an operator to return the parachute to the starting point.
- Requires tremendous amount of land in relation to the amount of work it performs.
- Does not easily adjust to changes in wind direction while operating.
- Does not operate high above ground where stronger winds will produce more power.
- Not well suited to residential applications or generation of electricity.

This analysis is representative of the type of thinking required prior to purchase of a WECS. Oddly enough, the particular illustration used here could be appropriate technology in communities where oxen are used in a similar fashion. Replacement of the oxen with a simple wind device, such as the parachute, would release the animal for other chores.

A more familiar illustration is shown in Figure 5-3. A conventional farm windmill supported by a tower drives a piston well pump by means of a vertical power shaft. The pump is mounted at ground level. This system is

64

used to supply domestic, stock, and irrigation water. Other wind machines are illustrated in Figures 5-4, 5-5, and 5-6.

How can wind machines be so different, yet one type can be best for some uses, and another type best for other uses? To answer this, we must examine their differences in some detail and develop ways of comparing different wind machines.

Figure 5-3: Farm windmill.

Figure 5-4: Savonius rotor.

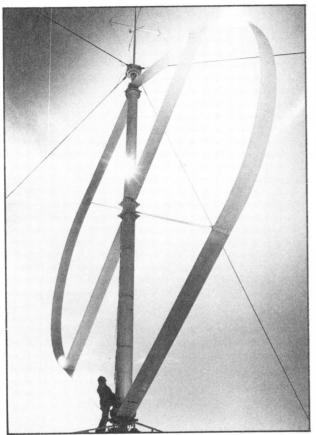

Figure 5-5: Darrieus "egg-beater" rotor.

Figure 5-6: Propeller-type wind turbine generator.

WIND SYSTEM PERFORMANCE

Comparing Different Types of Wind Machines

Figure 5-3 shows a farm windmill, and Figure 5-4, a Savonious rotor. While these two types of wind systems are very different, both windwheels present a large surface area to the wind in relation to the width and height on the machine. Notice that almost the entire disk area of the farm windmill is covered by blade surface; this presents a solid appearance to the wind. The appropriate term for this is *solidity*, which is the ratio of blade or windwheel surface area to *rotor swept area*, the area inside the perimeter of the spinning blades. Thus, solidity for the two machines illustrated in Figures 5-3 and 5-4 is nearly 1.0.

Solidity = blade surface area ÷ windwheel-swept area.

To calculate windwheel swept area, look at Figures 5-7 and 5-8. Swept area for a vertical-axis machine like the Savonious rotor is simply height times width. Swept area for disk-shaped windwheels is calculated from:

$$A = \pi \times D^2 \div 4, \qquad \pi = 3.14$$

A = swept area, in square feet or square meters, and
D = diameter in feet or meters

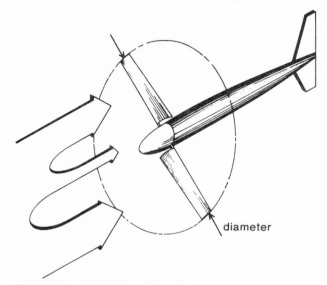

Figure 5-7: Horizontal-axis wind turbine.

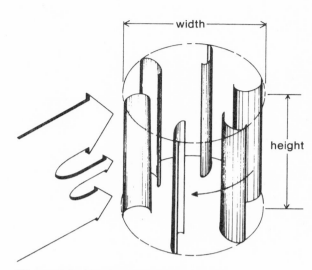

Figure 5-8: Vertical-axis wind turbine.

For example, the swept area of a 16-foot diameter windwheel is calculated as follows:

$$A = 3.14 \times 16 \times 16 \div 4 = 200.9 \text{ square feet.}$$

Mechanical drive applications, such as pumping water, demand very high starting torque* from the windwheel. The pump may have a load of water it is trying to lift from a deep well at the same time the windwheel is starting to turn. Further, high rpm operation (high revolution rates from the windwheel) is not required because it is generally better to pump a large quantity of water slowly than it is to pump a small quantity rapidly. This reduces resistance to water flow in the pipes. Larger-diameter, slower-moving pumps require slow-turning, high-torque windwheels, such as shown in Figure 5–3.

Electric generators operate by moving magnets past coils of wire. Two methods are available to get the required power from a generator:

- Large coils and strong magnets
- High-speed motion of magnet past coil

Most generators are actually a balance of these two design methods. However, to get, say, 2 kW out of a generator that turns at 200 rpm, the large magnets and coils might weigh as much as 300 pounds (135 kilograms). The same 2 kW can be generated by a smaller generator, which weighs about 50 pounds (22.5 kg), by spinning that generator at about 2000 rpm.

From this, we can see that a lightweight, low-cost wind turbine requires a fast-turning windwheel with much lower solidity, as shown in Figure 5–6.

One further relationship is needed to complete the discussion of solidity: *tip speed ratio*, the speed at which the windwheel perimeter is moving di-

---

*See discussion on torque in Chapter 2, p. 13.

vided by the wind speed. If the wind is blowing at, say, 20 mph (9 m/s), and a windwheel is turning so that the outer tip of the blade is also moving at 20 mph around its circular path, tip-speed ratio equals 1. Windwheels such as that in Figure 5–3 operate at tip speed ratios of about 1.

Suppose the tip were moving at 200 mph. With a wind speed of 20 mph the tip-speed ratio would equal 10. Low-solidity windwheels operate at tip speed ratios much greater than 1, usually between 5 and 10. We can now see a relationship between tip-speed ratio, which is a measure of rpm, and solidity. High-solidity windwheels spin slowly compared to low-solidity windwheels.

Figure 5–9 shows how the relative torque of various wind turbines decreases with increasing tip-speed ratio. As we noted previously, high torque requires a high solidity, and that type of wind turbine works best at low tip-speed ratios. Figure 5–10 shows how the best operating tip-speed ratio changes with solidity.

A wide variety of wind machines is sketched in Figures 5–11 and 5–12. Lest you get the immediate impression that there are more types of wind systems available than you might care to choose from, be assured that many of the types shown are only interesting historically. Some of the others are presently the subject of advanced concept studies.

The relative efficiencies of the types of wind turbines in which you might have an interest are illustrated in Figure 5–13. Notice that the efficiency is also related to tip-speed ratio, as is starting torque. As stated in Chapter 2, the maximum amount of power a simple windwheel (without a shroud or tip

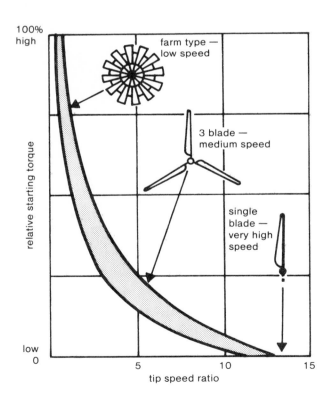

Figure 5-9: Relative starting torque.

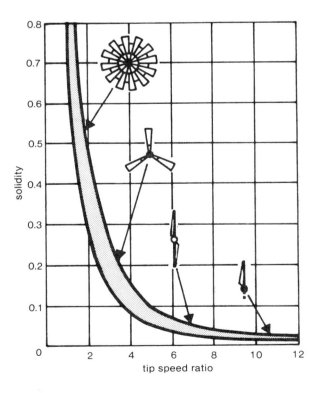

Figure 5-10: Solidity of several wind machines.

vanes) can extract from the wind is 59.3 percent of the wind power that would pass through that windwheel. From Figure 5-13, you can see that no windwheel actually extracts 59.3 percent.

Solidity affects design appearance in its relation to the number of blades a

Figure 5-11: Horizontal-axis wind machines.

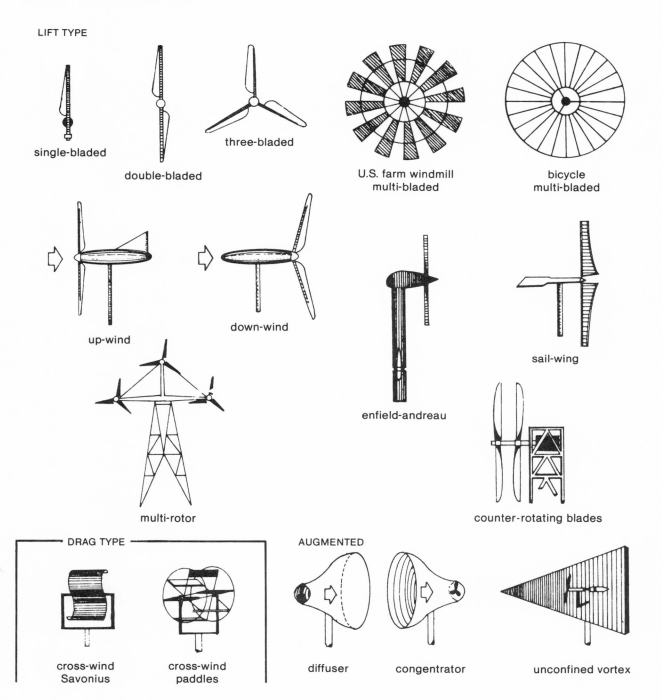

LIFT TYPE

single-bladed

double-bladed

three-bladed

U.S. farm windmill
multi-bladed

bicycle
multi-bladed

up-wind

down-wind

enfield-andreau

sail-wing

multi-rotor

counter-rotating blades

DRAG TYPE

cross-wind
Savonius

cross-wind
paddles

AUGMENTED

diffuser

congentrator

unconfined vortex

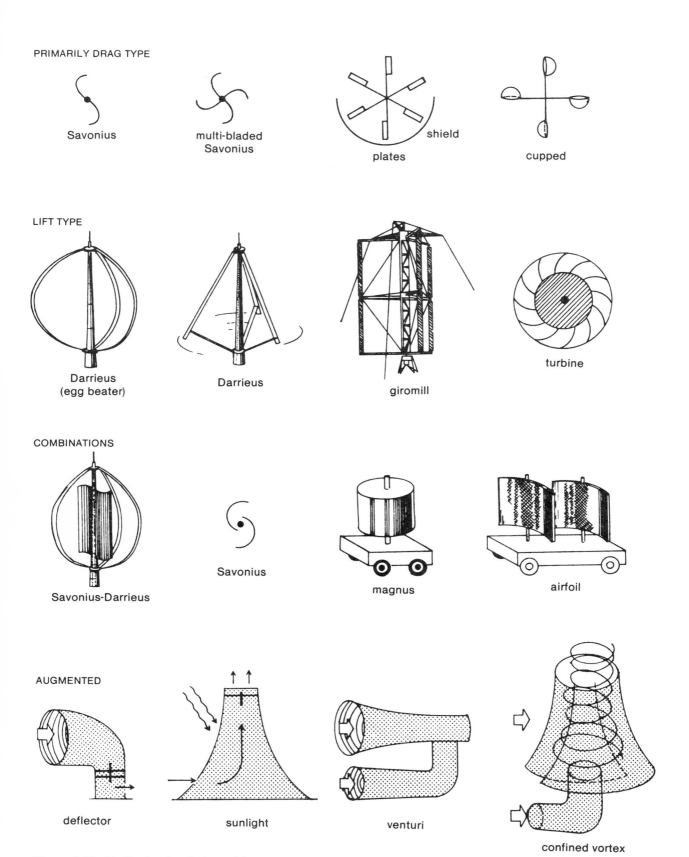

PRIMARILY DRAG TYPE

Savonius

multi-bladed
Savonius

plates   shield

cupped

LIFT TYPE

Darrieus
(egg beater)

Darrieus

giromill

turbine

COMBINATIONS

Savonius-Darrieus

Savonius

magnus

airfoil

AUGMENTED

deflector

sunlight

venturi

confined vortex

Figure 5-12: Vertical-axis wind machines.

70

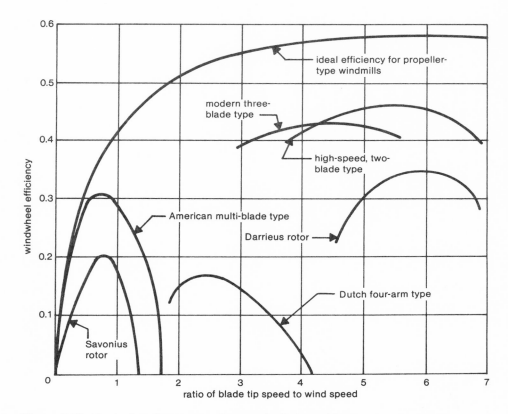

Figure 5-13: Typical performance of several wind machines.

machine has. High-solidity wind turbines have many blades; low-solidity machines have few, usually four or less. Figure 5–14 illustrates a wind turbine of intermediate solidity used for electric power generation.

There is more to a wind turbine than the solidity of the blades, the torque, the efficiency, or the load the windwheel drives. By looking at Figures 5–14 and 5–15 you can see two distinct methods for controlling the direction of the propeller-type machine: (1) upwind blades with a tail fin and (2) downwind blades that use the drag effect of the windwheel to keep the machine aimed directly into the wind. A third method, which works on either upwind or downwind-mounted blades, is a small wind turbine on the side of a downwind-blade machine (called a *fan tail*) that senses changing wind direction and drives the larger turbine around until it is aimed directly into the wind. As you might guess, the vertical-shaft machines like the Savonius rotor are always aimed into any wind direction.

Many of the wind turbines we have illustrated are designed with three or more blades. Two blades are occasionally used, but small two-bladed wind turbines usually need a larger tail fin than an equivalent three-bladed machine, or special weights to make the windwheel behave as a four-bladed unit. Small two-bladed machines exhibit a choppy motion in *yaw* (aiming into the wind). This is due to the natural resistance of spinning blades to changes in direction—something like a wobbling gyroscope.

Figure 5–16 illustrates a small, two-bladed machine that has governor-

Figure 5-14: Upwind rotor of intermediate solidity with tail fin control.

Figure 5-15: Downwind rotor.

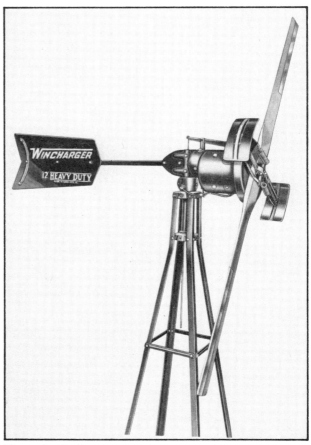

Figure 5-16: Small wind turbine generator with two upwind blades.

control mechanism weights in a position where another set of blades would otherwise be installed. For small machines, this approach is practical. For larger two-bladed machines, yaw (aiming) controls such as fan tails are more appropriate.

Wind System Power and Energy Calculations

As a first estimate, Figures 5–17A or 5–17B can be used to estimate the power output of any wind turbine in any wind. These curves use an overall efficiency for the rotor, transmission, and generator of 30 percent, which is a typical value for wind turbine generators. For example, let's assume the blade diameter is 20 feet (6 m) and the wind speed is 10 mph (4.5 m/s). From Figure 5–17B, this results in approximately 500 watts for a typical wind turbine. To convert watts to horsepower, divide by 746; so hp = 0.67.

72

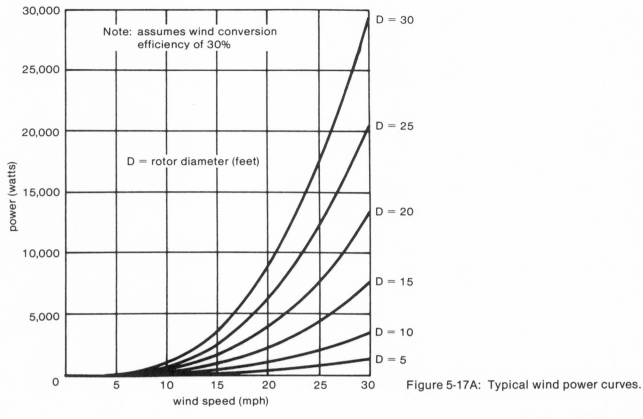

Figure 5-17A: Typical wind power curves.

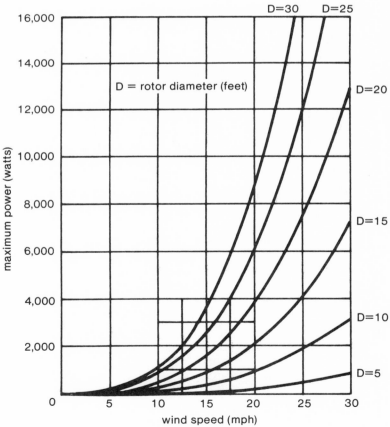

Figure 5-17B: Typical wind power curves.

73

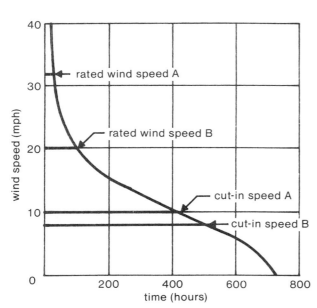

Figure 5-18: Power curves for two example wind turbine generators.

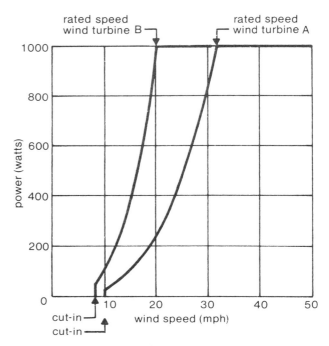

Figure 5-19: Example wind duration curve.

Manufacturers of wind turbine generators will supply sales literature containing power curves similar to Figure 5–18, with which you can make a more accurate determination of power and energy. To evaluate any wind turbine for its power and energy yields, it is important to consider the rated wind speed of the machine. This is the wind speed at which rated power is achieved. Also, you should know the cut-in speed, which is the wind speed at which the generator begins to produce power.

Figure 5–18 illustrates the characteristics and power curves of two hypothetical wind turbine generators of the 1000-watt size. Notice that unit A is rated at 32 mph, while unit B is rated at 20 mph. You can expect that wind turbine B is considerably larger in diameter than unit A.

As an example of the comparison of energy yields, Figure 5–19 is a hypothetical wind duration curve for one month. By dividing the 720 hours of that month into 20-hour segments and finding (on the graph) the average wind speed for each 20-hour segment, we can estimate the energy yield of each of the two wind turbines, using Figure 5–20*. Note that we are showing

---

*You can plot a wind duration curve from the average wind records of a nearby weather station and apply this technique for calculating wind energy if you desire. As an example the "Percentage Frequency of Wind" Table for Amchitka Island (Figure 3–22) can be used. Along the top line are wind speed categories, and along the bottom line, the percentage time the wind was blowing in each of these categories. Also, the percentage of time there was no wind, 0.9 percent, is listed in the second line from the bottom. January, the month for this record has 31 days, or 744 hours. Starting with the fastest speed category ≧56, the number of accumulated hours is simply:

≧ 56	$744 \times 2.1 \div 100$	$= 15.6$
48–55	$744 \times 3.7 \div 100 + 15.6$	$= 43.1$
41–47	$744 \times 4.8 \div 100 + 43.1$	$= 78.9$
	etc.	

The numbers in the right column are then plotted to make a curve similar to that in Figure 5–19.

No.	V mph	WIND TURBINE A		WIND TURBINE B	
		Power watts	Watts × 20 Hrs.	Power	Watts × 20 Hrs.
1	40	1000	20,000	1000	20,000
2	35	1000	20,000	1000	20,000
3	26	650	13,000	1000	20,000
4	22	320	6,400	1000	20,000
5	20	250	5,000	1000	20,000
6	19	238	4,760	950	19,000
7	17.5	162	3,240	650	13,000
8	17	150	3,000	600	12,000
9	16.5	138	2,760	550	11,000
10	16	125	2,500	500	10,000
11	15.5	113	2,260	450	9,000
12	15	100	2,000	400	8,000
13	14.5	90	1,800	360	7,200
14	14	80	1,600	320	6,400
15	13.5	70	1,400	280	5,600
16	13	60	1,200	240	4,800
17	12.5	50	1,000	200	4,000
18	12	45	900	180	3,600
19	11.5	40	800	160	3,200
20	11	35	700	140	2,800
21	10.5	30	600	120	2,400
22	10	25	500	100	2,000
23	9.5	0	0	95	1,800
24	9	0	0	87	1,740
25	8.5	0		70	1,400
26	8			50	1,000
27	7.5			0	0
28	7				
29	6.5				
30	6				
31	5.5				
32	5				
33	4.5				
34	4				
35	3.5				
36	2				
Total watt hours			95,420 = 95.4 kWh		229,940 = 229.9 kWh

Figure 5-20: Monthly energy from wind turbines A and B for previous wind-distribution curve.

this calculation as an example of energy estimation. You may not have to perform this calculation because consultants and dealers in wind machines would supply the information.

All that is required to make such a chart is to write down for each 20-hour section of the curve the average wind speed and the watts of power at that speed for each wind turbine (from Figures 5-19 and 5-18; you can use time intervals other than 20 hours if you wish). Then, multiply each power value times the 20-hour duration. This yields watt-hours. Add up all the watt-hours produced by each machine. Convert to kWh by dividing by 1000. In this example, wind turbine B yields roughly 230 kWh to wind turbine A's 95 kWh.

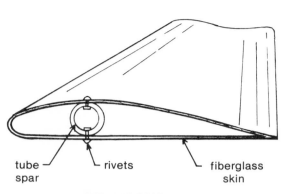

tube spar — rivets — fiberglass skin

**TUBULAR SPAR,
WITH MOLDED FIBERGLASS SKIN**

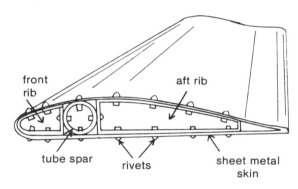

front rib — aft rib

tube spar — rivets — sheet metal skin

**TUBULAR SPAR,
WITH METAL RIBS AND SKIN**

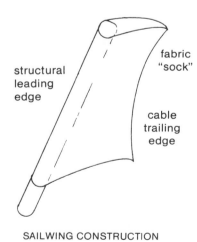

structural leading edge — fabric "sock" — cable trailing edge

SAILWING CONSTRUCTION

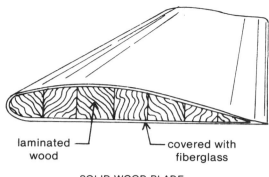

laminated wood — covered with fiberglass

SOLID WOOD BLADE

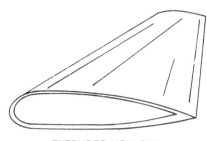

**EXTRUDED HOLLOW
METAL BLADES**

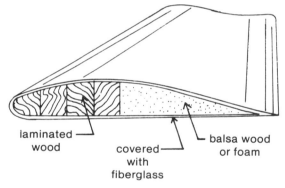

laminated wood — covered with fiberglass — balsa wood or foam

PARTIALLY SOLID BLADE

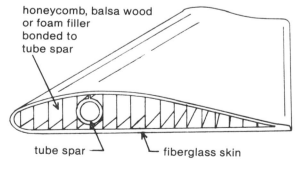

honeycomb, balsa wood or foam filler bonded to tube spar

tube spar — fiberglass skin

**COMPOSITE BLADE
CONSTRUCTION**

Figure 5-21: Different blade construction methods.

As indicated, a lower-rated speed implies a higher energy yield. If, for example, in the case just illustrated, all characteristics were the same, and wind turbine C is added to the comparison with a rated wind speed equal to the average wind velocity, which in this case is about 13 mph (value at 360 hours), the yield would be considerably greater. Wind turbine A is about 5 feet in diameter, wind turbine B is about 10 feet in diameter, while wind turbine C is about 19 feet in diameter.

You can expect the initial cost per kilowatt of rated power to increase with decreasing rated wind speed; at the same time yield (kWh) will increase, unless your wind distribution shows a considerable number of hours with wind speeds greater than 20 mph. For this reason, you need to know more than price and power rating. As you can see, rated wind speed is a valuable tool in wind turbine comparison.

WIND MACHINE ROTOR CONSTRUCTION

The major construction variations you find when selecting a wind turbine generally will involve the blades. The diagrams and discussion that follow are suited to propeller as well as Darrieus machines. One popular blade material is wood, either laminated or solid, with or without fiberglass coatings (Figure 5–21). Uncoated wooden blades usually have a copper or other metal leading edge cover for protection against erosion by sand, rain, and other environmental factors.

The extruded hollow aluminum blade first was installed on a WECS in the early 1950s. This blade construction is being used again, especially for the "eggbeater" Darrieus machines (Figure 5–5).

Built-up fiberglass blades with honeycomb or foam cores, or hollow cores except for a tubular structural spar, are also being used, as are built-up sheet aluminum blades. All these methods of construction have a history of service life in WECS applications as well as in many aerospace applications.

Rotor Speed Control   It is important to understand the various methods of rotor speed control. Blades are designed to withstand a certain centrifugal force and a certain wind load. The centrifugal force tends to exert a pull on the blades, whereas wind loads tend to bend the blades (Figure 5–22). A control is needed to

Figure 5-22: Loads on a wind turbine blade.

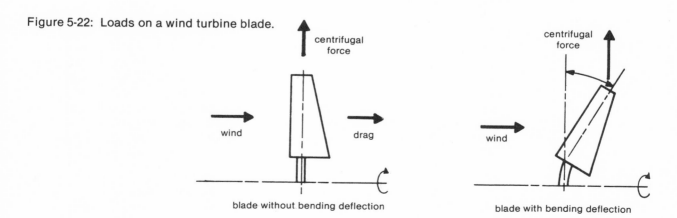

blade without bending deflection

blade with bending deflection

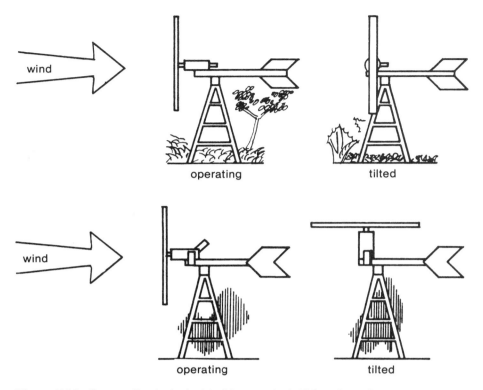

Figure 5-23: One method of wind turbine control: tilting the rotor.

prevent overstressing the machine in high winds. Obviously, one could design a wind turbine strong enough to withstand the highest possible wind, but this would be an expensive installation compared to a more fragile unit having a good control system.

Two primary methods exist for controlling a wind turbine: (1) tilting the windwheel out of excessive winds (Figure 5–23) and (2) changing the blade angles (feathering) to lower their loads.

Figure 5–23 (top right) shows a farm windmill in the shut-off position.

A method of control that was used extensively in Nebraska* is illustrated in Figure 5–24. Here, a wind fence is raised to block wind from flowing through the rotor. Appropriate for the type of windmill illustrated, this method has fallen into disuse with the invention of more sophisticated mechanisms. The Greeks, long before folks moved to Nebraska, controlled their sail windmills by taking in or letting out sail cloth. They knew when to take in sail because a whistle, mounted at the tip of one blade, would emit a loud noise whenever the machine was turning too fast.

Figure 5–25 illustrates a simple mechanical mechanism used to control blade angle (sometimes called *blade pitch angle*) for feathering the blades. Notice that the leading edge of the blade in its normal position is at an angle suitable to cause blade motion in the direction indicated. As blade rpm

---

*Erwin Hinckley Barbour, *The Homemade Windmills of Nebraska* (1899) reprinted by Farallones Institute, 15290 Coleman Valley Road, Occidental, Ca 95465.

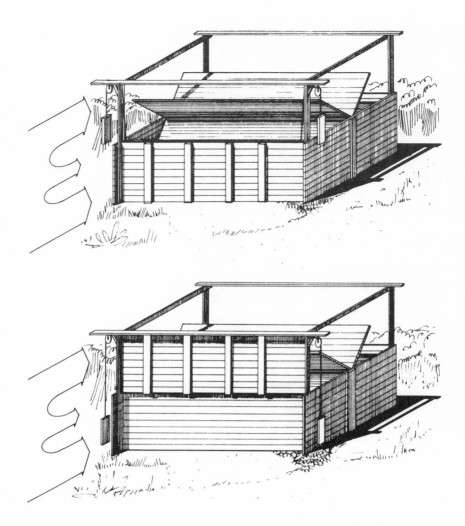

Figure 5-24: Control by wind blocking.

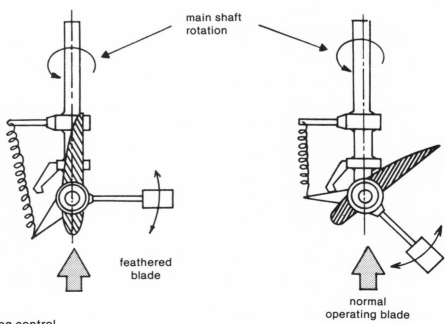

main shaft rotation

feathered blade

normal operating blade

Figure 5-25: Blade feathering control.

79

Figure 5-26: Icing on feathered blades.

Figure 5-27: Darrieus rotor with straight blades.

increases, centrifugal forces on the flyweight cause the weight, which is connected to the blade, to move around the blade center pivot shaft and cause the blade to pitch toward the feathered position. The feathered position pulls the leading edge of the blade into the wind to reduce or eliminate its driving force.

Ground control for shut-off or reset function can be combined with any of the design types to provide manual shut down of the windmill if required, such as during icing conditions, as illustrated in Figure 5-26. Other methods of blade control, such as automatic drag spoilers and hydraulic brakes, can also be used.

Ground control for the vertical-axis machines (such as Savonious type, Figure 5-4) can be accomplished by methods such as blocking the machine from the wind, as was used on old machines (Figure 5-24), by venting the S-shaped vanes, or by changing the S-shape to reduce torque.

The Darrieus rotor (Figure 5-5), which is commercially available in the "hoop" (turning rope) shape*, can be controlled using any one of two main methods:

1. A unique aerodynamic characteristic of the Darrieus rotor permits the blades to stall; that is, quit lifting or pulling, which slows it down. Stall is caused by overloading the blades with the generator, which requires an electronic control system.
2. Drag spoilers mounted either on the blades or the support structure can be mechanically or electronically actuated to slow the speed.

*This shape has been given the name *troposkein*.

80

Darrieus rotors designed with straight blades (Figure 5–27) are able to use the blade pitch control in addition to these methods. The particular Darrieus machine illustrated uses variable blade pitch to control the rotational speed.

Furling Speed

Another wind turbine characteristic you should consider is its *furling speed**—the wind speed at which the wind turbine is automatically or manually shut off. First, not all wind turbines have or need a furling speed. These machines are designed to survive while operating in even the most severe winds. For these machines, the manufacturer's specifications should state the highest speed for which the product is designed or has been tested.

Machines that must be furled will do so by automatic mechanical or electronic control, or you will be required to operate a ground shut-off control when high winds are anticipated. On modern machines, the tail vane may be locked sideways to turn the machine out of the wind, a brake is locked, or blades are locked in the feathered position.

Thus, the type of control system used in a particular wind turbine will play an important part in the furling operation.

ELECTRIC POWER GENERATION

As stated before, wind turbine generators tend toward low-solidity designs that have high tip speed ratios for high rpm operation. Before rural electrification, some of the early wind turbines used direct-drive generators. This means that the windwheel directly turned the generator, which was a heavy, low-speed unit. Many of the newer machines have transmissions mounted between the windwheel and the generator to increase the windwheel rpm by a factor called *gear ratio*—typically four, five, or more. Thus, for example, a windwheel speed of 100 rpm is increased by the transmission to 400 rpm or more. This allows for a lighter, lower-cost generator but requires the additional weight, cost, and maintenance of a transmission.

Lower cost is not the only factor enhanced by low weight. Heavier units can be more difficult to hoist onto the tower than equivalent lighter units. However, a factor in favor of direct drive is the lower maintenance required by elimination of the transmission. You can see that the choice between direct drive or a transmission drive involves many trade-offs.

Wind turbines have been built most often with the generator and blades mounted at the top of the tower. It is possible that machines will eventually have drive shafts at the bottom of the tower where the generator will be mounted. Another possibility is to mount a hydraulic pump at the windwheel and use hydraulic pressure to cause fluid flow through tubes down the tower to a hydraulic motor. With this concept, it becomes possible to have several windwheels powering one central generator.

Generators and Regulators

Generators installed in WECS can be of the direct current (dc) type or alternating current (ac) type. Many good books about generators are available in libraries, so a detailed discussion of the inner complexities of each

---

*Sometimes referred to as *cut-out speed*.

type of generator is not presented here. However, it is useful to understand the basics.

Alternating Current. Alternating current is generated in an ac generator (or alternator) by passing coils near alternate poles of magnets. The ac current generated is fed directly to the wires outside the unit. Here, you should understand that the *frequency* of the ac current is governed by the rpm of the generator (utility power is 60 hz [cycles per second] throughout the United States). The faster the coils of wire pass the magnet poles, the higher the frequency. To establish a fixed or constant frequency, the windwheel rpm must be held constant, regardless of wind speed. For small WECS, as considered here, the blade control device required to hold a constant rpm can be an expensive mechanism.

Generators used to produce ac power (which can be electrically fed directly into your load at the same time the utility ac power is wired in) are called *synchronous generators*. A wind turbine generator operating in a synchronous mode would be generating ac power at exactly the same frequency and voltage as the utility mains, or other source of ac power. This type of operation, as mentioned, increases the complexity of blade control systems and therefore increases the cost of the wind machine. On large wind turbines, synchronous generation is often practical.

Special generators are being developed that produce a constant frequency but allow for variable rpm by electronic compensation. These generators are being tested by several organizations, and the results of these projects might possibly allow turbine synchronous operation without the limitation of constant rpm (expensive) blade control systems. This type of generator is called a *field modulated generator.* Another method for generating fixed-frequency ac is to generate dc, and change the dc into ac by means of an inverter. Some inverters are designed to permit synchronous operation; others are not. We shall discuss inverters shortly.

Direct Current. Generation of dc usually involves generation of ac inside the generator, then conversion of the ac to dc by means of brushes and a commutator. This has been the common design for dc generators until recently. A method more commonly used now, because of improvements in diode technology, is to rectify the ac output of the alternator to dc. This technique eliminates the need for brushes and a commutator and takes advantage of the superior low-speed characteristics of the alternator over those of the dc generator. Virtually all new automobiles have the diode-alternator combination in their electrical systems. Thus, the ac power generated internally is fed to the battery and other loads as dc. Direct current is the only type of current that can be stored in a battery. Variable-frequency ac, as generated with a small wind generator, can be used without diode rectification to dc for many applications such as electric resistance heating. The current flow for each of these three generators is diagrammed in Figure 5-28.

Some experimenters have used the current before it is rectified to dc. Alternating current can be fed to a transformer, which steps up the voltage while lowering the current (amps). This reduces line loss (which we will discuss later) that can occur on long wire runs from a wind turbine to a load.

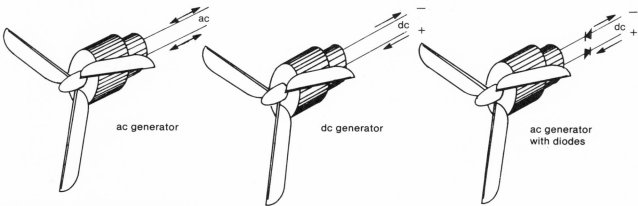

Figure 5-28: Three types of generators.

Thus, a 12-volt alternator is stepped up to, say, 100 volts. At the other end of the long run, another transformer (Figure 5–29) steps down the high voltage to an appropriate value, where the ac is rectified to dc. This method of transmission is subject to losses (about 5 percent) from the transformers. Transformers are designed for best operation at one frequency, while small WECS alternators generate variable frequency ac according to the rpm of the windwheel. Some transformers are designed for one frequency (60 hz or 400 hz), while others are designed to operate over a 50–400 hz range.

Direct-current generators have brushes made with carbon, graphite, or other materials to transfer electric power from rotating windings to the stationary case of the unit. These brushes transfer the full electric power of the generator. Some alternators have brushes also. In contrast to dc generators, however, these brushes transfer only the field current, a small percentage of the total alternator output. Alternators are also available without

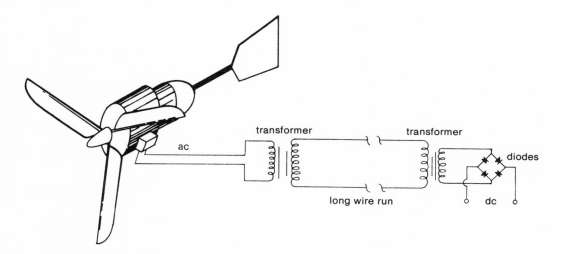

Figure 5-29: Using transformers and wind-generated ac for long wire runs.

brushes (brushless units) at a somewhat higher cost than equivalent brush-type units.

Some generators (or alternators) are available with permanent magnets. These magnets cause the electric current to flow as they spin past the coil windings. Other generators are available *field wound*, which means that electromagnetic coils requiring energizing current are installed in place of permanent magnets.

Regulating Output   Three methods are used for regulating or controlling the electric output of the generator:

1. Voltage regulators are used on field-wound units to control the strength of the field coils, which in turn controls the output voltage.
2. Voltage controllers may be used on permanent magnet units to adjust voltage levels according to the output of the generator and the needs of the system.
3. No regulation at all. The output of the permanent magnet generator is used as is, while that of the wound field is fed back to the field either directly, or through a resistor to give a variable-strength field according to the strength of the generator output.

Generators and alternators are selected or designed by WECS manufacturers according to their own criteria, which includes cost, weight, performance, and availability. Thus, in your own selection of a suitable wind turbine, you will find units with specially-made generators, as well as units with truck, automotive, and industrial alternators. You can make your own observations regarding availability of spare parts for custom, as well as industrial equipment. In some cases, the WECS manufacturer designs his own generator as a means of improving the overall system performance.

Figure 5–30 is a wiring diagram for a simple wind electric system. The wind turbine generator will charge the batteries which, in turn, will supply power to the two loads illustrated. Battery storage would be sized to store as much energy as is needed to make up for periods when the wind is lower than required or the power demand exceeds the wind generator capacity. The wind turbine would be sized to supply at least enough kilowatt-hours of energy as needed for the loads.

Suppose that the wind turbine generator supplies more kilowatt-hours than are needed. The batteries would be over-charged. Energy would be wasted. To preclude this situation, a load monitor is used (see Figure 5–31). The load monitor senses situations when the wind generator creates more power than the electric system needs, and reacts by switching on load C. Load C might be a resistance electric heater immersed in a water heater tank. It may be another battery bank, or any other load that will use the excess power. The load monitor thus prevents energy waste, and in so doing improves the energy utilization of the simple battery system.

Load monitors can be used another way. Suppose that the wind turbine generator does not supply the required energy. Perhaps a week of no wind occurs, and the batteries are nearly discharged. A load monitor can be used to sense this condition and activate a backup system.

The backup system could be a gasoline-powered generator, another set of batteries, or an extension cord to your neighbor's house. In any case, the load

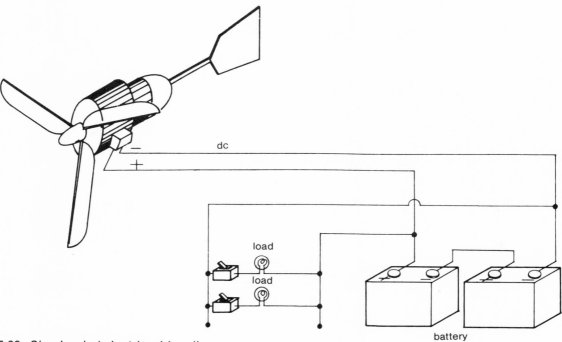

Figure 5-30: Simple wind-electric wiring diagram.

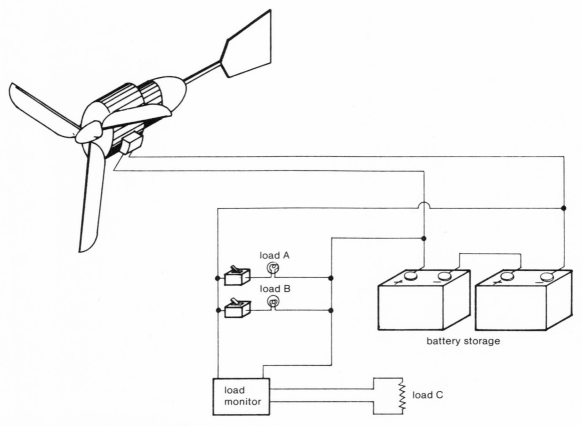

Figure 5-31: Wind-electrical system with load monitor.

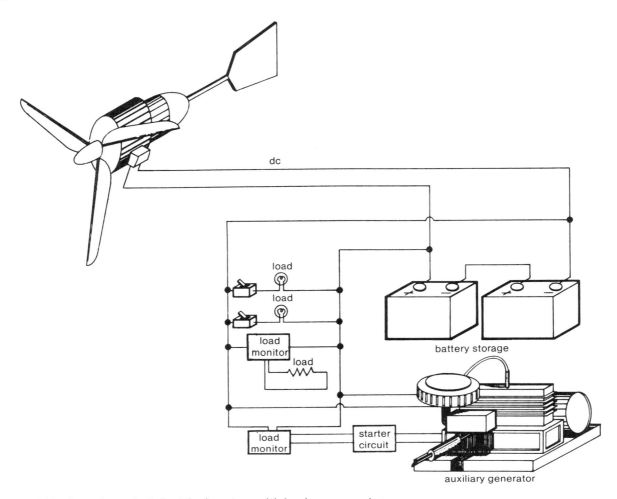

Figure 5-32: Complete wind-electrical system with backup generator.

monitor can control the source. In the case of the gasoline-powered generator, the load monitor can flash a light, ring a bell, or otherwise warn you of the situation, or it can energize the starter circuit on the auxiliary generator to bring it on line. Schematically, this could be done as shown in Figure 5–32.

ENERGY STORAGE

Batteries   The key to a viable wind electric system using battery storage is a low-cost, high-storage-efficiency battery. Figure 5–33 presents many of the important characteristics of three prominent types of batteries.

Lead-acid batteries of the automotive type are among the lowest-efficiency storage batteries available. Auto batteries usually retain about 40 to 50 percent of the energy your wind turbine generator will charge them with. Since it is energy storage you are buying, and dollars per kilowatt-hour of storage is the deciding factor, you should assume a low storage efficiency for such batteries when evaluating their usefulness.

Battery	Voltage per Cell	Density Watt-hrs/lb	Cycle Life	Storage Efficiency
Lead-Acid	2.0	10–20	200–2000	50–80%
Nickel-Iron (Edison Cell)	1.3–1.5	10–25	2000	60–80
Nickel-Cadium	1.2–1.5	10–20	2000	80

Figure 5-33: Characteristics of different types of batteries.

Golf cart batteries are one of the most suitable types available today, as are industrial batteries used for electric forklifts and units for standby power for computers, telephones, and electronic instrumentation. Standby batteries cost much more than golf cart batteries, but are designed for higher reliability and longer life. Golf cart and standby batteries are designed to be deep-cycled (discharged to very low charge levels) while auto batteries are not.

Batteries are rated by their voltage and by their storage capacity (amp-hours). For example, a typical golf cart battery might be a 6-volt unit, rated at 200 amp-hours. It is important to know that the amp-hour rating is based on a certain discharge rate. The typical rating is 20 hours for golf cart batteries. If 200 amp-hours were discharged over a period of 20 hours, the discharge rate would be 10 amps/hour (200 ÷ 20 = 10 amps). Greater discharge rates will result in a slightly reduced amp-hour capacity. You can get performance curves from battery manufacturers that illustrate this fact. You can convert the amp-hour and voltage ratings into watt-hours by simply multiplying the two. For example, 6 volts × 200 amp-hours = 1200 watt hours, or 1.2 kWh. With this, you can easily determine the number of batteries you will need.

> *Example 1:* Storage capacity needed is 30 kWh. If we use 6-volt batteries rated at 200 amp-hours, 30,000 watt-hours ÷ 6 volts = 30,000 ÷ 6 = 5000 amp-hours. With these 200 amp-hour batteries, we need 5000 ÷ 200 = 25 batteries connected in parallel. These batteries would be wired as shown in Figure 5–34.

Notice that connecting batteries in parallel increases the amp-hour rating of the entire battery bank (simply add up the total amp-hours available from each battery), while the output voltage remains the same as that of an individual battery. All batteries in a system must have the same voltage rating. An advantage to this arrangement is that any number of batteries can be taken away or added at any time to adjust your storage capacity to your needs.

> *Example 2:* Storage capacity needed is 10 kWh. For a 100-volt system which uses 2-volt batteries, it would require 100 ÷ 2 = 50 batteries wired in series to make 100 volts out of 2-volt cells. Each battery must have an amp-hour rating of 10,000 watt-hours ÷ 100 volts = 100 amp-hours. Thus, 50 2-volt, 100 amp-hour batteries satisfy the 10-kWh storage capacity requirement. These batteries would be wired in series as shown in Figure 5–35.

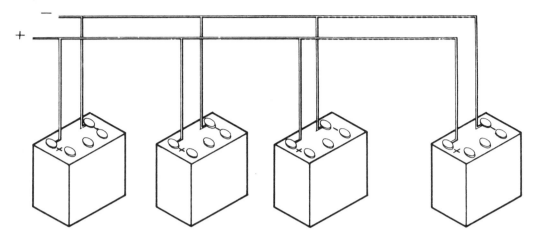

Figure 5-34: Battery storage bank—parallel wiring.

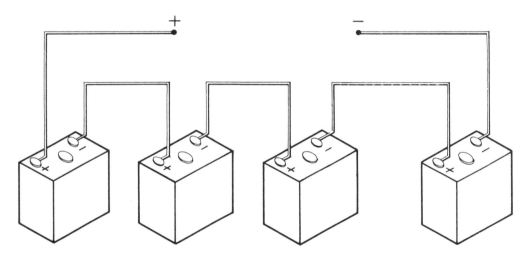

Figure 5-35: Battery storage bank—series wiring.

Notice that connecting batteries in series increases the voltage of the battery bank while the amp-hour rating of the bank remains the same as that of the smallest amp-hour rated battery in the bank. To increase the storage capacity of, say, a 100-volt battery bank, either increase the size of individual batteries or wire another 100-volt bank in parallel.

The cost of battery storage systems is relatively high. As a result of an increasing demand for electric personal transportation vehicles, a large amount of battery research has begun to reduce these costs. Research supported both by private and government funds is rapidly closing the gap between the batteries available on the market and potentially lower-cost, higher-efficiency units.

Figures like double and triple the energy density* are often quoted.

*Energy density is a measure of the amount of energy (kWh) per pound of battery. Since cost of items relates to weight of materials in them, higher energy density tends to enhance lower energy cost.

Batteries made with materials like nickel-zinc and exotic metals are being tested and developed to increase electric car performance. Wind electric system energy storage per dollar invested is also expected to improve.

Pumped Water Storage

It may be that you wish simply to store enough water for domestic uses, as discussed in Chapter 4. On the other hand, it may be that you prefer an electric system where wind power pumps water up a hill to a lake. The lake then supplies water stored with enough potential energy to operate a small hydroelectric system to recover the energy, as illustrated in Figure 5–36.

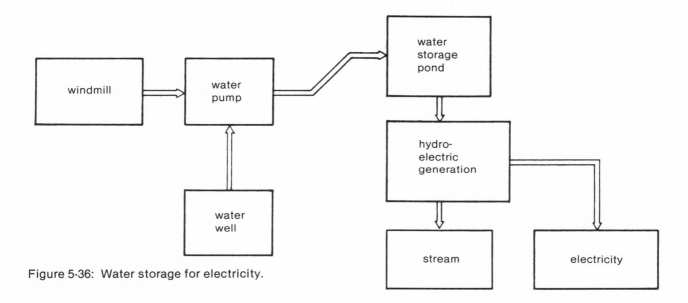

Figure 5-36: Water storage for electricity.

Figure 5–37 is a graph for determining the size of a pond or lake required to store a given amount of energy. Notice the large values. We sometimes hear of the idea that a few 55-gallon drums up in the attic ought to hold enough water to keep the lights burning all night, while the water trickles out of the drums turning a water turbine along its path. Unfortunately, this would not produce much power. For example, if you substitute a 1956 Oldsmobile for water and hoist the car 100 feet into the air (presumably between telephone poles, with an electric motor/generator winch driven as a motor for hoisting), you ought to get back only about 800 watts for ten minutes when you let the car fall under the control of the winch-turned generator.

The decision to use pumped water storage must be based on availability of land and such factors as: would the required amount of land be better used for something else, cost (bulldozing a lake can be expensive), and the end use of the pumped water. Electricity generation is just one use; irrigation, fire prevention, stock watering, fish farm, and recreation are others.

In the case of electric energy storage, a small hydroelectric system will probably exceed the cost of an equivalent battery system.

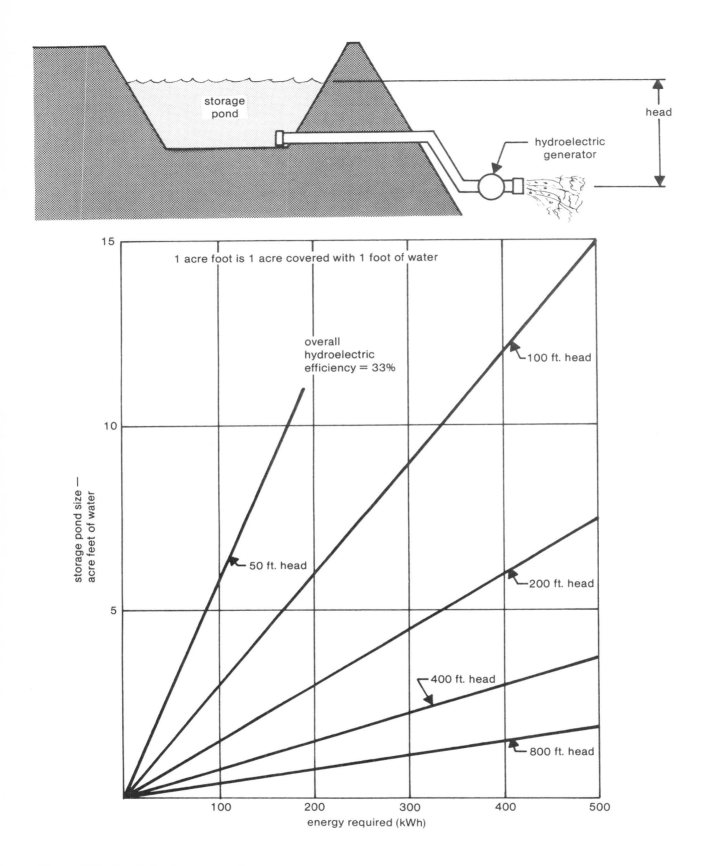

storage pond

head

hydroelectric generator

1 acre foot is 1 acre covered with 1 foot of water

overall hydroelectric efficiency = 33%

50 ft. head

100 ft. head

200 ft. head

400 ft. head

800 ft. head

storage pond size — acre feet of water

energy required (kWh)

Figure 5-37: Pond size for energy storage.

90

**Hot Water
or Hot Air Storage**
Wind power can be used to heat water for energy storage by splashing. Figure 5–38 schematically illustrates this method. Friction of water being pumped and splashed introduces all the energy as heat without the losses of a generator. As with any other heating system, heat loss can be kept to a minimum by insulating the tank. Splashing paddles can substitute for pumps. The cost effectiveness of this method has yet to be determined.

Water may easily be heated by wind electricity as diagrammed in Figures 5–39 and 5–40. In fact, this is perhaps the most efficient means of storing energy once the electricity has been generated. Where batteries are 60 to 80 percent efficient, electric heaters approach 100 percent. The only losses are those through poor insulation of the storage tank and electrical line losses.

Conversion factors of interest are: 1 kWh = 3414 Btu (British thermal unit); one Btu will increase the temperature of one pound of water by one degree Fahrenheit; water weighs 62.4 lb/ft³, or 8.4 lb/gallon.

Example: By neglecting heat loss through insulation, calculate the temperature rise in 30 gallons of water caused by 1 kWh of wind generated electricity.

30 gallons × 8.4 lb/gal = 252 lbs.
3414 Btu ÷ 252 lb = 13.5°F heat rise.

A wind-powered heating system is called a *wind furnace*. The heat supplied is best used for domestic or agricultural heating, as well as lower-temperature industrial applications. As illustrated in Figure 5–39, the wind-powered generator can be either ac or dc; regulated or unregulated. This provides some latitude in selection of wind turbine generator for a wind furnace, although manufacturers of the wind turbine, for many reasons, may not supply

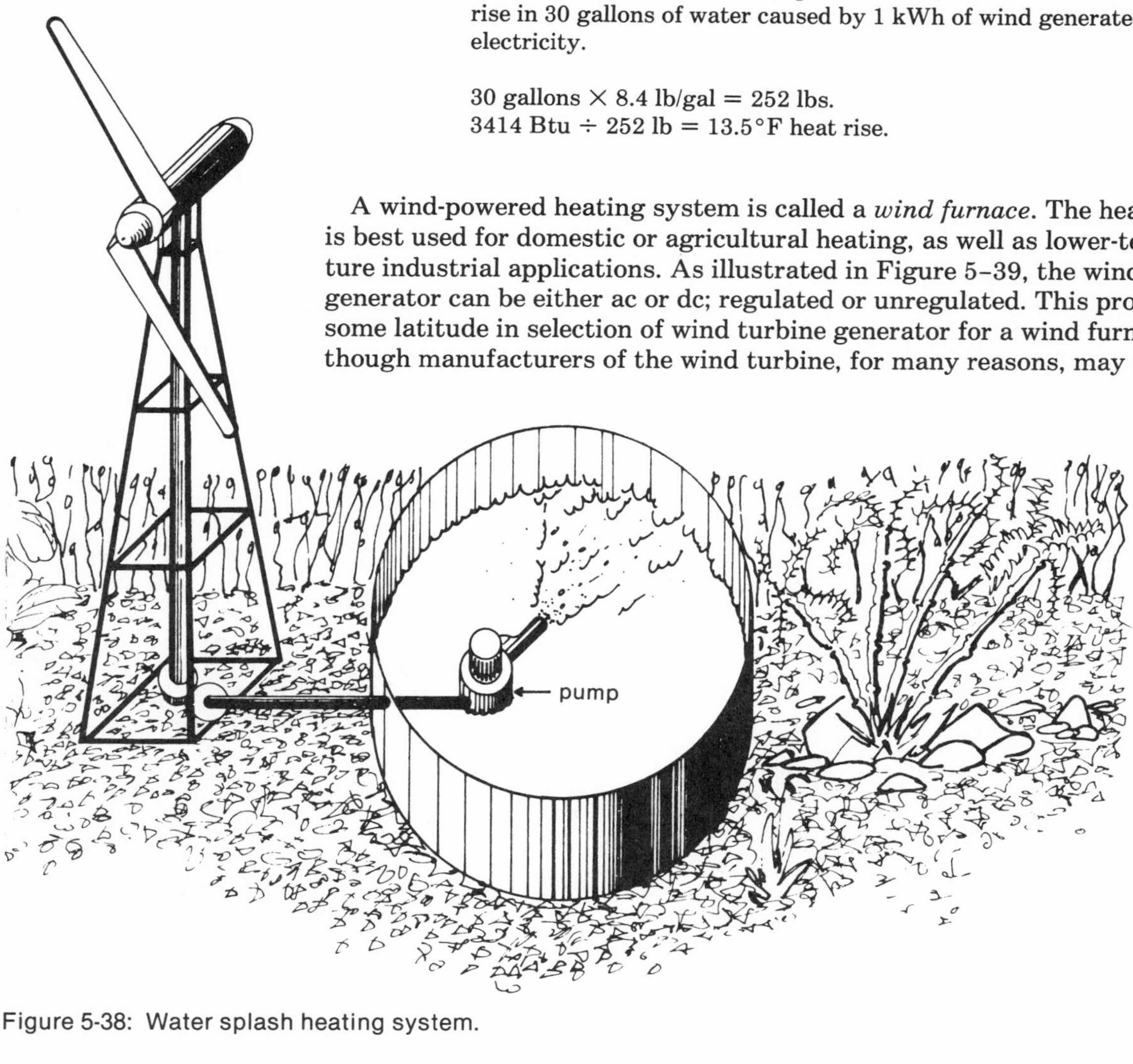

Figure 5-38: Water splash heating system.

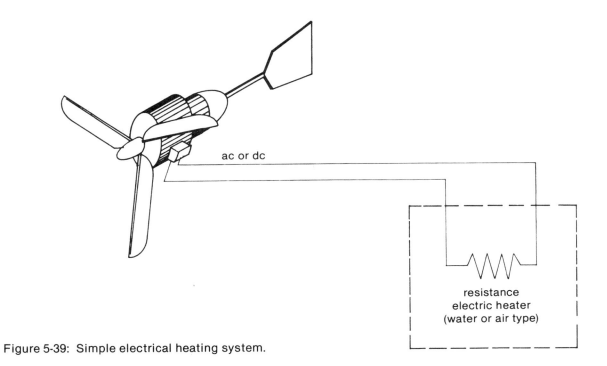

ac or dc

resistance
electric heater
(water or air type)

Figure 5-39: Simple electrical heating system.

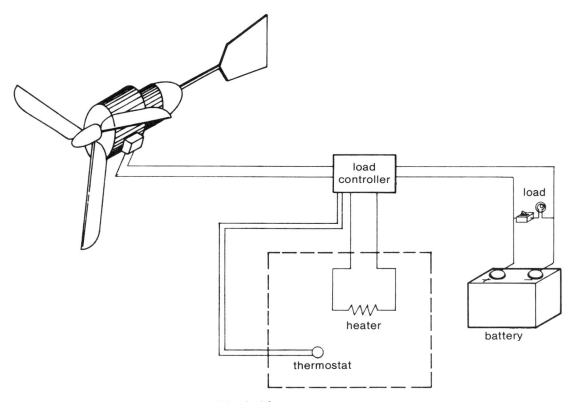

load
controller

load

heater

battery

thermostat

Figure 5-40: Complete electrical system with wind furnace.

all of the options. For them, it is more efficient to manufacture just one or a few types of machines that will serve the greatest number of applications.

The resistance electrical heater unit can be the *air* type, as with baseboard electric home heaters, or the *water* (or other liquid) *immersion* type. By using the immersion type, energy storage is provided by the thermal mass of the liquid, while thermal mass of the room (concrete floors, tile or brick walls, etc.) provides energy storage for the air heater.

If a regulated dc generator is used, then the wind furnace could look something like Figure 5–39, or Figure 5–40. A temperature monitor (thermostat) provides a control input to a load controller, which provides priority power to the heater, and secondarily, after the heater is warm enough, power to other loads.

Flywheels   Between 1951 and 1969, the Oerlikon Electrogyro bus was operated in Switzerland and the Belgian Congo. This was a flywheel-powered bus experiment in which a large flywheel was recharged at each bus stop.

Scientific American magazine (December 1973) discusses an improved flywheel energy storage device, the super flywheel. The super flywheel is different from the traditional spoke-rim flywheel, which propelled mine cars and elevators. The older gyros were large, heavy-rimmed wheels, like a tractor tire filled with water, while super flywheels are thin and tapered and spin at a high rpm. See Figure 5–41.

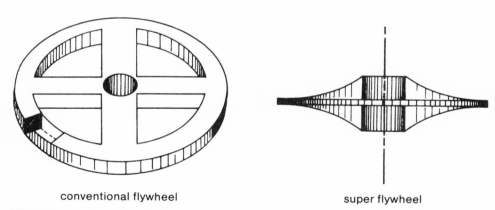

conventional flywheel   super flywheel

Figure 5-41: Flywheel diagrams.

To reduce losses in a super flywheel, the high-speed disc spins in a vacuum chamber, and precision ball bearings, magnetic bearings, or air bearings are used. These measures also enhance the safety of the flywheel by providing a housing that prevents damage, should the wheel fracture or fail.

Various design studies illustrate that a small, super-flywheel energy-storage unit should be cost competitive with an equivalent battery system and will most probably weigh less. Thus, this storage unit should see great potential in the commuter car applications, as well as small WECS. As this is written, no small super flywheel is commercially available, and few are being tested. Research is expanding in this area, however, and a super-flywheel energy-storage package could become available.

Synchronous Inversion— Energy Storage in the Utility Company Grid

Synchronous inversion allows power generation in phase with a utility network. Using a device called a *synchronous inverter*, wind-generated dc current is changed into ac current that has the same frequency as utility power and is fed directly into your house along with the current from the power company. As we discussed earlier in this chapter, another way to feed wind power into a utility system is with a synchronous generator.

Synchronous inverters have been used for years in applications such as regenerative drives for elevators. Here, the elevator is powered by the utility lines during its upward travel. During the downward travel, the motor becomes a generator that returns a portion of the upward power to the utility grid through a synchronous inverter while the elevator descends.

The immediate reaction to this idea is, "The electric meter would run backwards!" It would. There are three cases to note:

1. Wind generator not operating or not producing sufficient power for the load. Utility meter runs forward—its normal direction.
2. Wind generator supplying just enough power for the load. Utility meter stops.
3. Wind generator supplying surplus of power. Extra power will be fed to the grid. This will cause the meter to run backward.

The next response is usually, "But will my utility company allow this?" The answer to this question varies from state to state, and from utility company to utility company, and is rapidly changing.* The key question is development of a fair billing procedure for a complex problem (Chapter 7). Other questions have been or are being resolved, like safety, power quality and power factors. Regenerative drives such as electric elevators have been used for years, and the addition of WECS applications should not pose insurmountable technical problems.

---

*In effect, the Public Utility Regulatory Policies Act of 1978 requires utility companies to purchase excess power generated by wind turbines, if the turbine owners wish to sell this power. For additional information, contact: Federal Energy Regulatory Commission, Division of Public Information, 825 North Capital Street, N.E., Room 1000, Washington, D.C., 20426. (202) 357-8055.

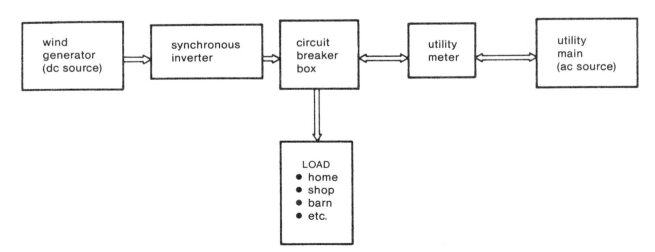

Figure 5-42: Synchronous inverter electrical system.

Of great importance to some WECS installations is that utility mains can be replaced in the diagram (Figure 5–42) by an ac generator powered by gasoline or diesel fuel. Certain technical details must be observed (the generator must have a higher power capacity than the wind generator), but the synchronous inverter will run well with an ac generator as the source of frequency and voltage. This is because the synchronous inverter uses the mains, or ac generator, as a reference for conversion of dc to ac.

Synchronous inverters are discussed here under energy storage because, in effect, you are using the utility grid as a storage cell for your excess energy. With the gasoline or diesel ac generator instead of the grid, energy "storage" results in effect from less fuel burned by the generator.

WIND SYSTEM TOWERS

As discussed in Chapters 2 and 3, it is often best to support your wind system, be it vertical axis or horizontal axis, to capture the higher winds above the ground and to be well above the nearest trees. Supporting a wind turbine that weighs several hundred pounds is no simple task and requires a rigid structure of some sort. Towers are subjected to two types of loads, as illustrated in Figure 5–43: *weight*, which compresses the tower downward, and *drag*, which tries to bend the tower downwind.

Towers are made in two basic configurations: guy-wire supported, and

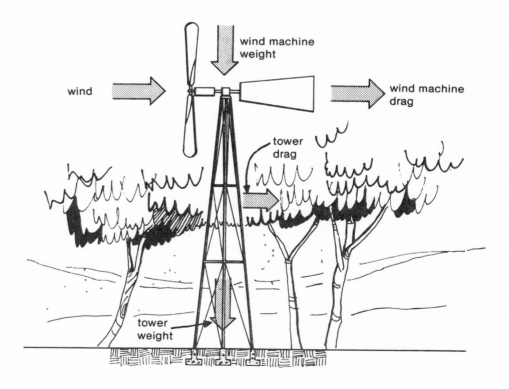

Figure 5-43: Tower loads.

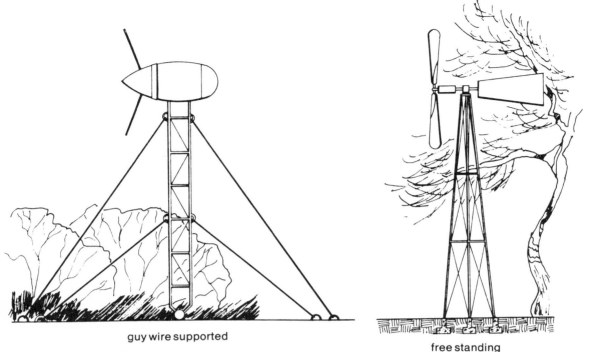

guy wire supported

free standing

Figure 5-44: Two types of towers.

cantilever or unsupported (sometimes called freestanding) (Figure 5–44). Also, given these two basic structural support configurations, towers are made with telephone poles, pipes (Figure 5–45), or other single-column structures, as well as lattice frameworks of pipe or wooden boards.

Regardless of which tower design you select, the overriding consideration is the selection of a tower that can support your wind turbine in the highest wind possible at your site. Of course, cost must also be considered. The load that causes many towers to fail is a combination of wind turbine and tower drag. A pair of skinny windmill blades may not look like they could cause much drag, but when extracting full power at rated speed they create nearly the drag of a solid disk the diameter of the rotor.

The ways in which towers typically fail are:

1. Freestanding towers buckle because of higher-than-design, wind-drag load from the windwheel.
2. A footing that anchors the tower to the ground becomes uprooted.
3. A bolt somewhere along the tower fails due to improper tightening (or falls out because not tightened at all), resulting in a tower weak point that eventually fails.
4. Guy-wire-braced towers buckle from improper spacing of the wires up the tower. Here, a tower that requires, say, three sets of cables spaced evenly along the length of the tower, gets only two sets, resulting in intercable spans greater than design specifications.
5. Guy wires fail from improper wire size, improper tension fasteners, or damage.
6. Guy-wire anchors uproot from the ground or come away from the structure to which they are attached.

When selecting a tower, consider the difference between guy-wire braced and freestanding. Here, you must know something about the structure and

96

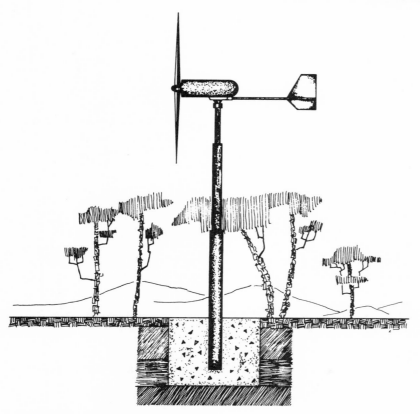

Figure 5-45: Freestanding pipe tower.

about the soil supporting the tower. A wise decision would be to consult your county agent concerning the type of soil you have and its ability to act as a foundation.

Building codes for your area will detail the basis for foundation design, and the wind turbine dealer or manufacturer should have drag data for the product you select. A registered professional structural engineer can perform any calculations necessary to ensure that a particular tower installation will support your windmill. The cost of professional services in the area of tower selection and design should be considered cheap insurance for a sound installation. Vibratory loads induced by the wind or wind machine also should be considered and professional advice may be required.

One other point to consider is the potential hazard of guy wires, particularly to playing children.

OTHER EQUIPMENT

Inverters   Inverters are devices that convert the dc voltage (power) to ac power. It is an age-old question whether dc or ac electricity is better. We do not propose an answer but suggest that many appliances are not designed to run on dc. Most motors, some stereos, TV sets, and certain other devices usually require 110 volts ac. Many wind-turbine generators are rated at 12, 24, 32, or maybe 110 volts dc. To change dc to ac requires an inverter.

Some inverters use a dc electric motor to drive an ac generator. By driving the generator at constant rpm, a constant frequency, usually 60 hertz (cycles per second) is generated. These units are called *rotary* inverters. Other inverters, called *static* inverters, are solid state, using transistors to switch dc into ac. Some units, the lower-cost variety, generate a square wave output, while more expensive units and rotary inverters generate sine wave outputs. The more desirable output is sine wave, especially if a stereo or TV set is to be operated. Square waves powering a stereo sometimes distort the sound.

Inverters are rated by their maximum continuous capacity in watts. A small surge capability is possible for most models. Thus, a typical 500-watt continuous transistor inverter might be rated to 700 watts for 10 seconds or even one minute, depending on the unit. Surge capability is needed, especially for inverters operating motorized appliances like refrigerators, because electric motors require considerable extra power for starting.

Selection of a suitable inverter involves another important factor, the efficiency of the inverter and, in more expensive models, automatic power adjustment. With a low-cost inverter, as would be available in most recreation-vehicle supply stores, the inverter will draw (from the battery) almost the maximum rated power, regardless of the load the inverter is driving. Thus, a typical 500-watt inverter may draw 400 or more watts from the storage system, while only powering one 100-watt light bulb. Higher-cost inverters offer the important option of a load monitor, which automatically adjusts the current drawn by the inverter, according to the load.

A typical efficiency curve for a static inverter looks like Figure 5-46. From this, you can see that wherever possible it is best to select inverters that will operate near their maximum rated capacity. This could mean using several small inverters for various loads or one large automatic inverter for the entire system. In any case, the cost of such inverters may dictate which inverter is selected. Ultimately, the cost of energy supplied is the primary consideration. Costs are discussed in Chapter 6.

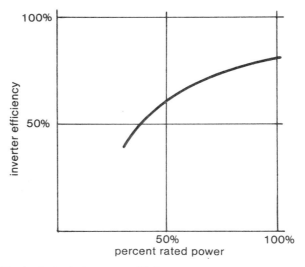

Figure 5-46: Typical static inverter efficiency.

**Backup
Equipment**

We have already mentioned backup options during previous discussions. The concept of backup implies that wind is the primary energy source and that something else is the backup. That something else could be the utility lines (switched off until the wind dies down), solar heat, a gasoline or diesel generator, solar cells, or an extension cord to your neighbor's house.

Auxiliary engine-generator sets that burn gasoline, propane, or diesel oil are readily available from equipment supply houses, catalog sales stores, and dealers of wind equipment. You might consider recharging your wind system batteries by jumper-cabling your automotive electrical system to your batteries. As you warm up the car in the morning before driving to work, charge your battery bank. This is not a suggested practical alternative to a more conventional approach to backup power, but in a pinch it will work. Power take-off (PTO) generators are available for tractors. These units will recharge batteries or otherwise provide backup power. Selection criteria for any backup equipment would include maximum power requirements, cost, reliability and maintenance, and environmental factors such as noise and exhaust. Auxiliary equipment can be purchased with automatic controls or can be installed with the requirement of user control. It is a good idea to exercise your auxiliary equipment on a regular basis (even when not needed), to ensure it remains in proper operational condition.

CHAPTER 6

Selecting Your Wind Energy Conversion System and Figuring the Cost of Its Power

It is surprising how many WECS installations have been bought only on the basis of "first cost" rather than satisfaction of energy requirements. Often a WECS owner will choose a system on a basis such as: "This system will supply about 60 percent of my energy needs and only costs $3,400." The information in this chapter should help guide you in determining the best system for your needs, the resulting actual cost of power, and the likely cost of power from the utility company in the future.

"First cost" and emotional factors such as habits and desires often have a strong effect on one's estimate of energy requirements. "First cost" of a wind system has more than once convinced a family moving to a remote location that two television sets operating off a wind-charged battery bank is an unacceptably high level of luxury.

THINKING OUT YOUR MOST APPROPRIATE SYSTEM

Figure 6-1 can focus your thinking on the steps required to accomplish a rational selection of the most appropriate WECS.

Your demands will establish whether the windwheel will drive mechanical devices, such as pumps, compressors, or grinding wheels; or electrical devices, such as generators or alternators. Mechanical devices demand a windwheel design of relatively high solidity (Figure 5-3), whereas electrical generators, for reasons mentioned earlier, tend to be equipped with relatively low-solidity windwheels.

Once established, the type of windwheel work performed enables you to evaluate the devices on which the work is performed; such as pumps and generators.

To visualize some of the many practical ways a windmill can be used as part of a complete system, study Figure 6-2. You can follow any path on this diagram that leads from top to bottom, from wind turbine to user. You will

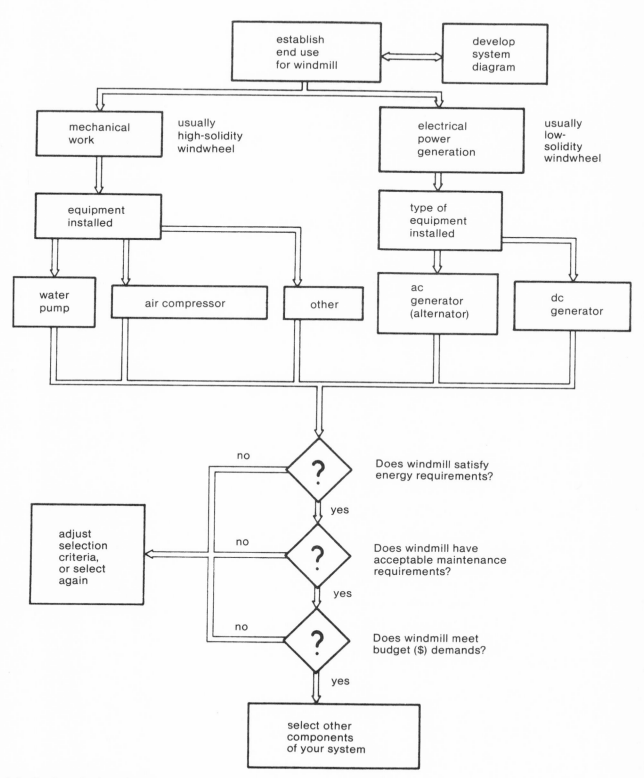

Figure 6-1: System design steps.

Figure 6-2: WECS choices.

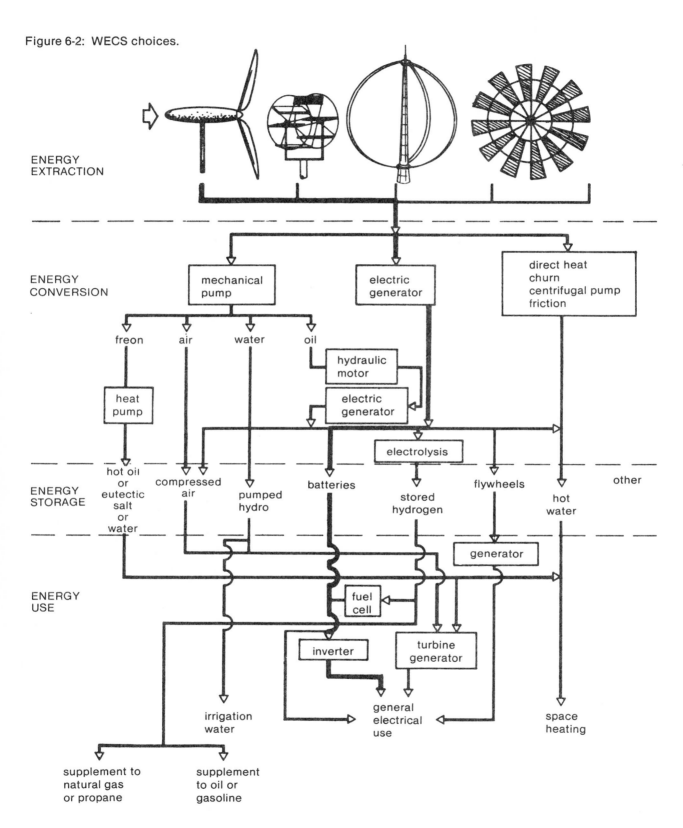

ENERGY EXTRACTION

ENERGY CONVERSION

mechanical pump

electric generator

direct heat
churn
centrifugal pump
friction

freon   air   water   oil

hydraulic motor

electric generator

heat pump

electrolysis

ENERGY STORAGE

hot oil
or
eutectic
salt
or
water

compressed
air

pumped
hydro

batteries

stored
hydrogen

flywheels

hot
water

other

generator

ENERGY USE

fuel cell

inverter

turbine generator

irrigation
water

general
electrical
use

space
heating

supplement to
natural gas
or propane

supplement
to oil or
gasoline

102

note most of the practical energy devices and processes along each path. The diagram illustrates the most common system, as well as some systems being developed. A bold line connects the components in a common wind-energy system.

A basic system consists of the energy source (wind), a conversion device (wind turbine), energy storage (batteries, pumped water, etc.) energy use (such as heaters, motors, TV set), and a backup source of energy (such as gasoline generators or solar cells).

Some of the options you have in selecting the best system to pump water are: a wind turbine mounted directly over the pump using a push rod; a drive shaft to a remotely located pump; an electric pump system; a hydraulic or pneumatic system (Figure 6-3); or a jet pump geared directly to the wind turbine rotor (not illustrated).

Figure 6-4 is a chart that presents many of the factors you would consider in the selection of the best water pumping system for your needs. You would normally evaluate each option by following the blocks of Figure 6-1. It may be that the question block regarding cost and your budget will eliminate several of the options from your list.

Another example of widely different choices available for accomplishing a

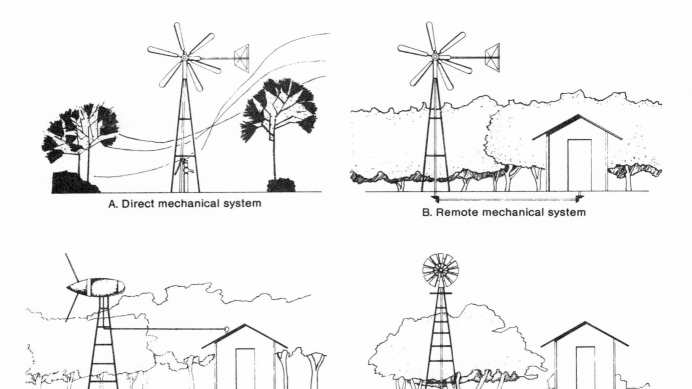

A. Direct mechanical system

B. Remote mechanical system

C. Wind electric system

D. Hydraulic/pneumatic system

Figure 6-3: Water pumping options.

	METHOD	ADVANTAGES	DISADVANTAGES
A	Windmill direct over well-driving pump at well with drive shaft or push rod	Simple Possibly relatively low cost Equipment has long history of availability	Well site may not be a good wind site
B	Drive shaft to remote well site	Allows some flexibility in WECS siting Allows power take off for other requirements	Safety hazard of drive shaft Relatively high cost of drive components
C	Wind generator electric pump	Allows energy storage (in batteries, etc.) Electricity for other requirements Relatively high efficiency Equipment has long history of availability Allows best flexibility in WECS siting	Energy loss in long wire runs Relatively high cost Safety requirements of electric wire runs
D	Windmill—hydraulic/pneumatic system Water can be pumped directly by bubbled air, or by pneumatic pump	Allows greater flexibility of WECS siting than B Hydraulic power or compressed air available for other requirements	Energy loss in long fluid pipes Safety considerations of compressed fluids or air Relatively high cost
E	Jet pump geared directly to windmill Two or three pipes come down the tower from the pump (not shown)	Allows greater flexibility of WECS siting than B Common, relatively inexpensive pump is relatively efficient Avoids generator and motor losses of the electrical system	Minimum wind speed for developing minimum head for pumping may be quite high Must prime the pump at the top of the tower

Figure 6-4: Different possible ways to pump water with a wind system.

specific task is depicted in Figure 6–5. Two methods are illustrated for preventing the freezing of a stock water pond and reducing fish kill caused by ice blocking the absorption of oxygen into the water. Either method works, and there are many other possibilities. One method uses a small Savonious rotor, which is mounted above the pond and drives a propeller that churns the water. This circulates warmer water to the surface, preventing ice formation and adding oxygen to the water. The air-pump method bubbles air into the water and also causes the warmer water to rise to the surface.

A COMPARISON OF POWER COSTS
USING TWO DIFFERENT WIND TURBINES

All too often a WECS installation is purchased according to first cost rather than performance. Since "The cheapest windmill will do!" is never really an appropriate criterion, you should understand what "the cheapest windmill" really is.

Analyze, as an example, a simple wind electric system that is used only for heating water (Figure 5–39). Compare the two hypothetical wind turbines that were described in Chapter 5 (Windmill Power and Energy Calculations). Both are rated at 1000 watts of power, and their power curves are shown in Figure 5–18. The wind-duration curve is shown in Figure 5–19. Unit A is about 5 feet in diameter and has a rated wind speed of 32 mph. It produces

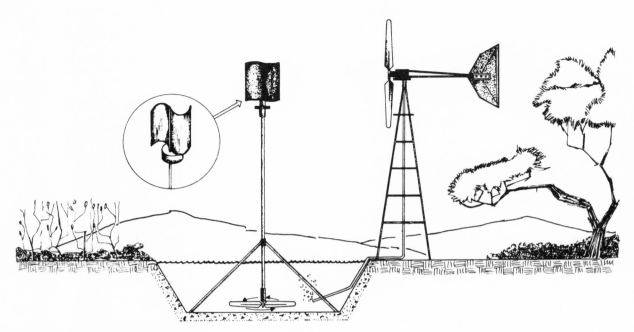

Figure 6-5: Two methods for wind-powered pond aeration.

95 kilowatt-hours in the month illustrated in Figure 5–20. Unit B is about 10 feet in diameter, has a 20 mph rated wind speed, and yields 230 kWh in the same month. For simplicity, assume you already have the necessary wire and electric water heater. Therefore, just consider the wind turbine and tower costs, plus installation. The following table shows these hypothetical cases.

Hypothetical Initial Costs of Two Wind Turbines

	Wind Turbine A	*Wind Turbine B*
Wind machine	$1,500	$2,000
Tower	$ 500	$ 600
Installation	$ 500	$ 600
TOTAL	$2,500	$3,200
Yield/year, kWh	1,140	2,760

Notice the difference in costs. Wind Turbine B requires a stronger tower and a somewhat higher installation cost.

If you selected solely by first cost, Wind Turbine A would be the obvious choice. However, Wind Turbine A will yield 1140 kWh/year at your site, while the more expensive Wind Turbine B will yield 2760 kWh/year.

Two more factors will be considered: maintenance cost and depreciation. Assume both machines depreciate fully to no resale value in 20 years. Also, assume that Wind Turbine A, the smaller of the two, costs $100 a year for maintenance, while Wind Turbine B costs $150 per year. Without considering any interest costs (covered later in this chapter), calculate the total cost of ownership over the whole 20-year life span:

	Wind Turbine A		Wind Turbine B	
First Cost		$2,500		$3,200
Maintenance	20 yrs. × $100 =	$2,000	20 yrs. × $150 =	$3,000
TOTAL COST		$4,500		$6,200

Total yield in 20 years and resulting cost per kilowatt-hour are:

Wind Turbine A	Wind Turbine B
1140 kWh/year × 20 years = 22,800 kWh	2760 kWh/year × 20 years = 55,200 kWh
Cost = $4,500 ÷ 22,800 = 19.7¢/kWh	Cost = $6,200 ÷ 55,200 = 11.2¢/kWh

Wind Turbine A costs 19.7¢/kWh, while Wind Turbine B costs 11.2¢/kWh. However, Wind Turbine A costs only $2500 to purchase and install, compared to $3200 for Wind Turbine B. If you ask which one is really cheaper, you must consider other factors. Wind Turbine B has roughly twice the energy yield of Wind Turbine A, but, from your calculation of energy requirement in Chapter 4, do you really need twice the yield? Will the extra yield allow future growth that you forgot to allow for in your energy requirement calculations? What about bank interest on the money you must borrow, or lost interest on the funds you take from savings to purchase your WECS? Do wind turbines really depreciate fully to no resale value?

Before continuing, note here that if you are presently paying 6¢/kWh from a utility company and this cost increases at 7 percent per year, at the end of 20 years you will be paying 22¢/kWh, and your average cost per kWh would have been 12.3¢ during that period (the way these numbers are calculated will be described later).

GATHERING THE FACTS
FOR YOUR ECONOMIC ANALYSIS

The analysis of economic factors can be as complex or as simple as you wish to make it. As in the above example, for a simple analysis you need to know the following:

1. Total installed cost (dollars)
2. Expected system life (years)
3. Total energy yield over the entire system life (kWh, hp-hr, etc.)
4. Annual maintenance and repair costs (dollars)
5. Other annual costs and savings (dollars)
6. Expected resale value at end of service life (dollars)
7. Other factors

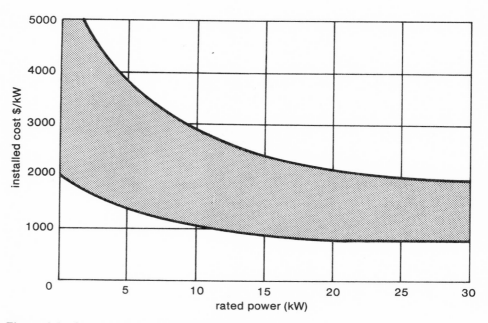

Figure 6-6: Costs of complete wind electric systems.

**Total
Installed Cost**

Generally, the bigger a WECS system is, the less it will cost per unit of rated output. The installed costs of WECS, measured in dollars per kilowatt rated power, tend to decrease with increasing rated power (Figure 6–6).

This total cost is broken down into component costs (Figure 6–7). These pie charts show the relative costs by the size of the pie slice. For the wind electric system, batteries may cost as much as the wind turbine, as illustrated. The pie slices may not actually reflect the system you are planning, but it is a good idea to look at the relative costs.

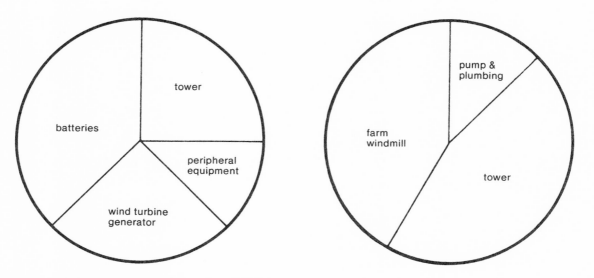

Figure 6-7: Typical relative costs for a small WECS.

Costs normally include everything you must purchase and install to provide normal or desired operation of the system. This would include:

1. Wind turbine
2. Tower, footing, guy wires, etc.
3. Batteries
4. Pumps
5. Storage sheds for batteries or other equipment
6. Storage tanks
7. Wires
8. Plumbing
9. All installation costs such as delivery, plumbing, electrical, and building permits

You may wish to read Appendix 2 to decide if you want to try to do your own installation or have it done for you.

When simply comparing system costs, as we did in the first example, some costs are often omitted, like the wiring and water-heater unit. These were assumed to be already available. Such omissions are not always valid, though, as in a case where the wind-electric water heater is being compared with a propane gas or solar-powered water heater. Here, water heater costs may be very different.

Expected System Life

Wind turbines have been installed at the South Pole and performed for more than 20 years. Wind turbines have been installed in the Rocky Mountains and have been destroyed, or damaged, in just a few months. Both locations are windy and both are subject to severe weather conditions. In trying to analyze the expected life of your system, you are confronted with several problems. No 20-year rating of wind turbines by some consumer-oriented organization is available. It would be nice to simply assume that the manufacturer's or dealer's statements (if any) concerning expected life are valid for your site. To do so will require intuition and an unemotional, scientific guess of the relative credibility of such statements. Most manufacturers tend to shy away from making claims on the life of their units.

Wind system designers, however, tend to plan their equipment for useful lifetimes of much longer than 20 years. Bearings, belts, and some parts may have to be replaced during such a time span, but the basic machinery—if well designed—should last. The U.S. Department of Energy Small Wind Systems Test Center at Rocky Flats near Golden, Colorado is testing designs and will publish results that will help answer this question. Eventually, dealers and manufacturers may publish such test data in their product literature.

The logical first estimate in any case comes from:

- Your wishes or needs
- Dealer's estimate
- Interviews and opinions of other WECS owners

Perhaps you intend to own a WECS for a limited time; use this time value in your cost study. Maybe a nearby WECS owner has had good service for 10 years; see if you can find out his expectations for continued service life (and his technique for getting such good service). Start with these and the dealer's comments. Probably you will not be far off.

Total Energy Yield

This value, expressed in kWh, or horsepower-hour (hp-hr) is the result of the energy resource study and site analysis you perform using the methods in Chapter 3. Chapter 8 illustrates how to match energy needs with energy availability. Total energy yield represents the work of your entire system planning process.

You will likely estimate or calculate energy requirements on a monthly basis. Simply add these together for all the months of a year, and multiply the annual total by the expected life. This results in the total energy yield you expect from your system.

Annual Maintenance and Repair Costs

It is possible to purchase a maintenance contract from some WECS dealers. Depending on the terms of such a contract it might be possible to use the cost of the contract, plus a small contingency cost for replacement of broken parts, as the annual maintenance cost.

Another approach is the "other owner interview." Find out what everybody else is paying to keep their machine operating. At the same time, try to evaluate the maintenance practices of the owner relative to manufacturer's recommendations. Some owners of water pumpers will report 20 or more years of good service and that the only maintenance performed was an occasional topping of transmission oil. One such performance was obtained from a machine whose manufacturer recommended an annual *oil change*! From these interviews and discussions with the dealers, form an estimate of the annual maintenance costs.

Expected Resale Value

Some farmers who bought wind electric systems prior to the advent of the Rural Electrification Act are reselling their old machines for prices varying from scrap-iron rates to the original price. Allowing for inflation, this would indicate that these WECS owners are enjoying as little as 50 percent depreciation in value over a 40-year life span.

Other folks sometimes purchase these machines, rebuild them, and resell them at prices reflecting inflation. An old Jacobs may have been purchased for $900, sold 25 years later for between $100 and $1000, repaired and restored to its original condition, and resold for $2000 to $3000. History shows that wind turbines, if properly maintained, can sell for their original cost plus an addition for inflation.

The resale price of old machines has risen rapidly in recent years. Greatly increased demand, coupled with the availability of old machines at reasonable cost has contributed to the resale value trend. Introduction of new wind turbines from more manufacturers could soften the resale price structure, but this depends upon the ability of the WECS industry to satisfy the demand. It seems reasonable to expect some resale value.

Other Annual Costs and Savings

Added to the list of costs are: the bank interest you pay for money you borrow to purchase a WECS (or money you do not earn if you withdraw from savings), insurance, and taxes.

Taxes. Taxes, as inevitable as the wind, work both ways. First, the bad news—your tax assessor may be delighted to see you erect that permanent-looking structure! Rather than bothering to figure the property tax rate, percentage of assessed value, homeowner's tax rebate, and all the other present-day gimmicks, you can calculate your tax rate by dividing your total annual real estate taxes by the true estimated value of your property. For instance, if your "spread" is worth $50,000 and you pay $500 in taxes, your tax rate is really $500 ÷ $50,000 = 0.01 = 1 percent. So, you would expect to pay about $30 tax on a $3,000 wind system.

Next think about some income tax angles. If your wind system is used for your farm or business, you can depreciate it a certain amount each year; that is, you include in your cost of doing business part of the original purchase price until it has been charged off to a preset salvage value. A reasonable lifetime for a wind turbine for tax purposes is expected to be 10 years for wind-electric types and 15 years for water pumpers. The salvage value at that time may be 10 percent of the original cost. Such values are always conservative—less than the actual life if the device receives reasonable maintenance and less than actual resale value. If only part of the energy or water produced is used for the farm or business and the rest is for personal use, depreciation may be applied only to the farm and business portions. If you sell the wind turbine for more than its depreciated value, the excess is taxed as capital gains. Note that the cost of utility power that is no longer needed for the farm or business is a lost expense item of your tax form. Finally, if you borrow money for your wind turbine, the interest is tax deductible.

As this is being written, new tax laws are being drawn up in various states and at the federal level that promise tax relief and may provide incentives which would reduce the total cost of WECS ownership. Your local congressperson can give you the details of any such legislation.

Insurance. Additional homeowner's insurance will be another added cost. You may not feel a need for fire insurance, but liability insurance is a must. The last section of Chapter 7 discusses some insurance problems. Depending on the cost, local windstorm conditions, and dealer's warranties, wind damage insurance, if offered, may be desirable.

Investment cost. The cost of your investment is a very important item to consider. If you plan to have a long-term loan on the wind turbine, you may select the lifetime of your wind turbine, for analysis purposes, to be the same as the bank loan (this does not have to be the 10 years used for tax purposes). The annual cost of your investment is then the same as your annual payments. This number leads to a more accurate way of calculating your costs than that used in the example at the beginning of the chapter. There, total maintenance costs, which are spread out over the life of the machine, were added to installation costs. This is like adding apples and oranges.

If you take money from your savings account to buy a wind turbine, you could just take the interest you lose on that money, minus the income tax you would have paid on the interest, as the cost of your investment. This is not the true rate, since you cannot count on anyone returning your investment to you at the end of the life of the wind turbine. Rates set up just like loan payments are the desired ones for a correct analysis—equal, regular payments that include both capital payback and interest. The following table gives these annual payback rates for various interest rates and lifetimes:

Interest Rate	Loan Period (years)					
	5	10	15	20	25	30
4%	$0.225	0.123	0.090	0.074	0.064	0.058
6%	0.237	0.140	0.103	0.087	0.078	0.073
8%	0.251	0.149	0.117	0.102	0.094	0.089
10%	0.264	0.163	0.132	0.118	0.110	0.106
12%	0.277	0.177	0.147	0.134	0.128	0.124

Annual loan payment per dollar borrowed

The most appropriate value for you to use is the one you have obtained on your invested money in the past, minus taxes. For example, using the Wind Turbine B described earlier with a 6 percent interest rate on your money for a 20-year life would mean a payback on the $3200 invested (no salvage value) of $279 per year ($3200 × 0.0872). Adding on the estimated annual maintenance charge of $150 gives an annual cost of $429 for 2760 kWh, so the new estimate for the cost of power is (without considering the tax angles) 15.5¢/kWh ($429 ÷ 2760).

Inflation. Other factors deserve consideration. Your estimation of inflation rates for utility power and wind systems can greatly influence your final decision on wind power. A 9 percent, long-term rate increase for utility power is one estimate at this time. How your electrical bill could increase (for the same power consumption) is indicated in the accompanying table for several rates of inflation. This table can be used for the price change of any other service or product, or annual interest on money.

Rate of Inflation	Value or Cost At:						
	1	5	10	15	20	25	30 years
3%	1.00	1.13	1.30	1.51	1.75	2.03	2.36
5%	1.00	1.22	1.55	1.98	2.53	3.23	4.12
7%	1.00	1.31	1.84	2.58	3.62	5.07	7.11
9%	1.00	1.41	2.17	3.34	5.14	7.91	12.17
11%	1.00	1.52	2.56	4.31	7.26	12.24	20.62

Effect of Inflation on Future Value, Such as Cost of Energy

This table shows, for instance, that a 9 percent annual inflation rate leads to a new value 5.14 times as much in 20 years. If your present electrical rate is 5 cents per kWh, it would then be 5 × 5.14 = 25.7 cents. If propane or heating oil costs 40 cents per gallon now, it would be $2.06 per gallon in 20 years.

The average cost of these items over the years, (instead of the final value) is shown in the accompanying table.

Rate of Inflation	Average Value or Cost Over:					
	5	10	15	20	25	30 years
5%	1.11	1.27	1.46	1.76	1.91	2.21
7%	1.15	1.42	1.79	2.31	2.53	3.15
9%	1.20	1.58	2.17	3.07	3.39	4.54

The average costs over 20 years for the electricity and propane in the table would be 12.8 cents and $1.02.

If occasional utility power outages require you to have an alternate energy supply, you may wish to compare the relative costs of a gasoline-driven portable or stationary power unit and a wind turbine generator. The initial cost of the wind turbine and storage system is probably considerably greater than the cost of the power unit, but the wind turbine might more than make up for this difference by producing usable power much of the time.

CHAPTER 7

Possible Legal Hurdles

This chapter deals with possible legal problems caused by purchasing, installing, and operating a wind energy conversion system (WECS). If you install a wind turbine within a few hundred feet of your property line, and your neighbor plants a row of fast-growing trees along that line, your wind energy could, in a few years, be greatly reduced. Protecting yourself from this occurrence is described in this chapter under "Wind Rights." A height limitation on structures in your local zoning ordinance is one of a number of possible problems that may affect your wind-turbine construction plans. The section "Obtaining a Building Permit" explains your courses of action. Your optimized wind-electric system may mean interconnection with the local utility company or sharing with your neighbors. Either can have economic or legal implications. (See the section "Sharing, Buying, and Selling Power"). Finally, before something goes wrong, warranty, liability, and insurance matters deserve your consideration.

Nearly all the material in this chapter has been taken from a comprehensive report compiled by George Washington University.* You and your attorney should refer to that report for more detailed information. The material presented here is for general information only. You should contact appropriate local and other governmental authorities if special permit problems are anticipated, or your attorney if easement or liability problems arise.

ANY RESTRICTIONS IN YOUR DEED?

Sometimes agreements have been made by the present or prior property owners of a potential wind turbine site not to conduct certain activities or erect buildings greater than certain heights on the property. These private agreements are commonly known as *restrictions* or *restrictive covenants*,

---

Legal-Institutional Implications of Wind Energy Conversion Systems (Washington, D.C.: George Washington University, 1977). Available from National Technical Information Service, U.S. Department of Commerce, Springfield, VA 22161.

some of which are said to "run with the land." Owners succeeding the person who entered into such an agreement are bound to comply with the restrictions. A title search of the deed should reveal any such agreement. However, these agreements must fulfill various legal requirements before a person can be bound by them. Therefore, their mere existence may not necessarily mean that a WECS owner is legally bound to follow them.

WIND RIGHTS AND THE "NEGATIVE EASEMENT"

There are no laws that describe your right to the wind that blows across your land. If wind turbines come into widespread use and conflicts arise, such laws might be enacted. The problem will be very real to you if your wind turbine is within a few hundred feet of an upwind neighbor who plants trees or builds a tall structure so that the smooth flow of your wind is interrupted, causing a reduction in the output of your wind turbine. Worse yet, if his structure is upwind during your strongest gale winds, the added turbulence might just be enough to destroy your blades.

The document that can provide the WECS user with the greatest protection is a *negative easement*. Easements are interests in another person's property that give the easement holder a limited right to use the property for a specific purpose, for example, a right-of-way. A negative easement gives the easement owner the power to prevent certain acts of an upwind landowner. Short of buying the property, this is the best way of protecting your wind source.

Consider trading partial negative wind easements with your neighbor. These easements would allow each of you to block each other's wind only by your own wind turbines, possibly specifying their diameter and minimum distance to the common property line. The natural growth of existing trees should probably be excluded. Obviously an attorney should be consulted.

OBTAINING A BUILDING PERMIT

After determining that you have a satisfactory site for a wind turbine and deciding that you want to erect one, investigate any possible laws or local ordinances that may affect the erection of your tower and wind machine. A call to your local building inspector may be all that is required.

The categories of controls that affect the wind system owner are: (1) local zoning ordinances, (2) federal, state, and local laws, and (3) building costs. Each of these types of controls is discussed below.

Zoning   Zoning regulations are based on the state's jurisdictional powers, under which the state may regulate private activity for the purposes of enhancing or protecting the "public health, safety, and welfare." Zoning laws, usually called *ordinances*, are, with a few exceptions, enacted and enforced by municipal and county governments. Where zoning is in force, the WECS owner must show that his proposed activity and structure conform to the

Figure 7-1: Negative easements can protect a wind system from neighboring trees.

restrictions applied to the site by zoning ordinances before obtaining a building permit.

The majority of municipalities and counties use the same basic process to enforce zoning ordinances. The prospective wind-turbine owner, or his contractor, starts the process with an application for a building permit, filed with the planning department, zoning enforcement office, building inspector, etc., who will issue such a permit if the provisions of the applicable zoning ordinance are met. Construction may be checked periodically to insure that the materials and workmanship meet the building codes.

The typical lattice-type windmill tower is generally termed an *accessory building* since it is a separate structure and cannot be lived in. If it is part of a residence or other building, the zoning restrictions are applied to the whole structure.

Zoning ordinances typically regulate uses or activities that may occur on the land; population density; and such building requirements as height, number of stories, size of building, percentage of the lot that may be occupied, and setback. Aesthetic considerations, though not typically treated as a separate concern, are a factor in writing and administering zoning ordinances. So-called "architectural review" is not standard practice, but some municipalities have enacted legislation designed specifically to regulate building appearance and compatibility with neighboring structures.

Conceivably, the proposed WECS may violate restrictions, particularly in a

115

residential area. Resolution of this problem may depend on the wording of the ordinance and its interpretation. Also, the WECS may be permitted as an accessory use; one related to a permitted use of the land. Under this theory, for instance, ham radio towers have been permitted on residential property.

There are various options available to the WECS owner who is restricted by a zoning ordinance. He can appeal the interpretation and application of the ordinance to the board of zoning appeals or board of zoning adjustment, a local body that exists to oversee the zoning process. He may be able to utilize the so-called special exception or conditional use – a permitted use that is explicitly mentioned in the ordinance but whose application to a particular area is allowed only after approval by the board of zoning adjustment.

He may also be able to get a *variance* – a permitted variation from the ordinance. The granting of variances is governed by broad considerations of the purposes to be served by the zoning scheme, and the board of zoning appeals is often the name of the body with this power. Next, the WECS owner might attempt an amendment to the ordinance prohibiting his operations. In most states, the procedure for amending ordinances is the same as that for enacting them in the first place. This usually involves action by the city council or board of supervisors of the municipality. Also, the WECS owner might attempt to get a change ordered by the courts. This, of course, involves a court action, which might proceed under a variety of special procedures. Such an action is likely to be expensive and time consuming and may succeed only in extreme circumstances.

Other Laws and Regulations Relevant to Land Use

Federal, state, and local regulations and laws, or statutes other than the zoning ordinance, conceivably may affect the WECS owner. These include statutes that regulate the selection of sites for electric generating plants; laws designed to protect the environment; and legislation regulating the use of particular geographic areas, such as coastal lands, wildlife reserves, historic sites, or navigable waters. It is very difficult to generalize about the impact of these provisions, but their overall effect on a small WECS should be minimal. Which laws affect the WECS owner will depend on the wind system's location and size and possibly who wants to erect it: an individual, a cooperative, or a company. If a permit from the federal, state, or local government is required before construction can begin, other laws may become involved in the permit process. For example, if the tower were located within the high-water mark of a river and a permit from the Corps of Engineers were required, federal laws require the Corps to consult other governmental agencies before issuing a permit.

In addition, legislation designed to protect the environment exists at both the federal and state levels. Most of these acts become relevant to the WECS owner only when a government permit-granting agency is involved. Generally, compliance with the law is the direct responsibility of that organization, not of the developer. This involves the satisfaction of various paperwork requirements and the submission of such reports to a variety of interested agencies. Overall, because of the likely low environmental impact of a small WECS, such procedures will probably be time consuming at worst. Further, the small WECS owner is not likely to be affected by a state's power plant site selection statute, since these typically apply to a minimum rated capacity of about 50 megawatts or to those utilizing a certain fuel source.

Federal Aviation Administration (FAA) regulations require that the owner of any structure higher than 200 feet give notice to the FAA on forms provided for that purpose. While few small WECS towers are that tall, lower height limitations apply within the vicinity of an airport. For example, a wind turbine up to 100 feet high (to the top of the blade path) might be allowed at 5,000 feet from a runway. As soon as notice (if necessary) is received, the FAA applies different height standards to determine whether the tower is an obstruction. These standards are generally less stringent than those governing notice. If the WECS were found to be an obstruction, the most likely requirement would be the placing of warning lights on it. Applications for building permits for structures in the vicinity of airports are usually forwarded to the FAA by municipalities.

Building Codes Like zoning matters, the state's jurisdictional power is the basic authority under which building codes are enacted. Some state legislatures enact statewide building codes while others delegate the authority to the local governments. One or a combination of the four model building codes has been adopted by most states or municipalities. These codes, and the geographic area dominated by their association/author, are:

1. The Uniform Building Code written by the International Conference of Building Officials (adopted primarily in the West)
2. The Basic Building Code compiled by the Building Officials and Code Administrators, International, Inc. (found in the Northeast and North Central areas)
3. The Southern Standard Building Code enacted by the Southern Building Code Conference (adopted in the South)
4. The National Building Code, developed by the National Board of Fire Underwriters

Local variations exist despite the model codes. Some municipalities have adopted selected provisions rather than the entire code. Interpretations of the same code differ from city to city.

Unlike zoning ordinances, most building codes apply retroactively. Three types of information are provided in most codes: definition of terms; licensing requirements; and standards. Taken together, the definitions and licensing requirements have the effect of prescribing who is authorized to conduct particular sorts of construction activity. For example, unless you are doing the work on your own system, the International Association of Plumbing and Mechanical Officials Code states that only licensed plumbers may do work defined as plumbing. Many codes require that structural design plans be prepared by a state-certified engineer.

Two types of code standards exist: technical specifications and performance standards. Codes prescribing technical specifications set out how, and with what materials, a building is to be constructed. Performance standards represent a more progressive and technically more flexible approach. Codes based on these standards state product requirements that do not prescribe designs and materials. For example, "the structural frame of all buildings, signs, tanks and other exposed structures shall be designed to resist the horizontal pressures due to wind in any direction . . ." Typical construction components specified in codes are structural and foundation loads and stresses, construction material, fireproofing, building height (this

Figure 7-2: An unusual wind generator site is this parking lot in northern Vermont.

represents a common duplication of the zoning ordinance), and electrical installation. The WECS developer is likely to be required to comply with the standards for structural and foundation loads and stresses, as well as the electrical installation code. The structural design standards set out the minimum force measure in pounds-per-square-inch that the WECS must bear under certain circumstances (e.g., wind or snow). The electrical code regulates the use of a generator and the electrical wiring when voltage levels are above 36 volts.

118

Administration of the building codes is delegated to a board of review in some states, and to the building official in others. No building may be erected, constructed, altered, repaired, moved, converted, or demolished without a building permit, and this can be obtained only after the building official is satisfied that the plans satisfy all applicable building codes. A trend is developing toward combining the administration of building codes and zoning ordinances in one municipal department.

Dissatisfaction with the building inspector's denial of a permit may result in an appeal before the local board of building appeals. The common bases of appeal provided by the codes are: an incorrect interpretation of the code by the building official; the availability of an equally good or better form of construction not specified in the code; and the existence of practical difficulties in carrying out the requirements of the code. The local board members are usually appointed experts in the field of construction. The local board may uphold, modify, or reverse the building official's decision. Further appeals to the state board of building appeals or to the courts are also available.

SHARING, BUYING, AND SELLING POWER

You may be considering a wind system where you share excess power or sell it to neighbors, buy makeup power, or sell excess power to a utility. Unfortunately, the state utility regulatory structure may cause you some unwanted problems.*

The first of these possible problems has to do with the regulated monopoly structure of the public utility industry. This structure operates by the assignment (based on what is often called the *certificate of public convenience and necessity*) of a geographical area to a particular electrical supplier. To operate within an area occupied by an existing utility all entities defined as public utilities typically must obtain this certificate, and to do this they usually must demonstrate a compelling need, such as the inadequacy of existing service. The result of this is often to prevent new electrical suppliers from operating within such a protected domain. Their status as a public utility is the crucial point here.

Selling Power to Neighbors   The small wind-turbine owner who generates power for only personal use or shares it with neighbors at no charge is not defined as a public utility and thus will not be hindered by the regulated monopoly structure. Sales of electricity to others, however, may cause problems. Some state statutes limit the exact number of people (i.e., 10 or 25) to whom sales can be made before public utility status exists. More commonly, the statements in the law will contain language making it appear that *any* sale to *any* part of the public will result in public utility status and the need for a certificate, effectively

---

*A new federal law, the Public Utility Regulatory Policies Act of 1978, in effect, requires utility companies to purchase excess power generated by wind turbines, if the turbine owners wish to sell this power. For additional information, contact: Federal Energy Regulatory Commission, Division of Public Information, 825 North Capital Street, N.E., Room 1000, Washington, D.C., 20426; (202) 357-8055.

prohibiting such sales. However, the courts of such states often interpret this language to require a "dedication to the public use"; that is, an offering of service to the general public coupled with a willingness to serve those who apply. Under such a standard, the small WECS owner-operator selling to a few friends would probably escape. Sales of electricity by a landlord to his tenants may cause similar problems, and the states have taken a variety of approaches here. For instance, if each tenant is metered and billed apart from the rent charged, public utility status may be hard to avoid. Finally, if the WECS owner obtains supplemental power from the existing utility grid, the service contract with the utility will almost certainly contain terms prohibiting such sales.

One possible way for small WECS owners to avoid these problems is for them to start a cooperative. Basically, co-ops are nonprofit, membership corporations, the members being both the owners of the corporation and the consumers of the electricity produced by it. Most co-ops are fairly large, located in rural areas, and funded by the Rural Electrification Administration. However, most states have special statutory schemes for the incorporation of co-ops, and these often allow incorporation by as few as three to five individuals. The various requirements, of course, must be compiled with by the WECS owners.

The point of utilizing the cooperative form in this context is that cooperatives are not defined as public utilities in some states, although the number of states that make this exception is decreasing. In states that do make this exception, the WECS owners could generate power for themselves within the domain of a regulated utility without being checked by the certification requirement. However, in such a case they would not be granted protection from competition with existing or future utilities.

Generally, this problem of restriction of WECS operations due to collisions with the existing regulated utility structure may be more hypothetical than real, at least for sales to a very few people. However, given the diversity of laws and practices in the fifty states, this may not always be true. Experts in the field or perhaps the state public utility commission should be contacted before such sales are attempted. It should be remembered that the more substantial the sales to others, the greater the likelihood of problems.

Buying Power

At present, utilities generally do not object to user-owned power generation systems that provide the user's power needs part of the time while the utility provides the power when the system is turned off. However, increased use of solar and wind-power generating systems may bring about a change in attitude of the utilities toward these systems due to the fact that heavy demands could be placed on the utility during windless or sunless periods. The need for the utilities to maintain peak load capacity, even though it would generally be selling less electricity, would result in a loss of revenues. Conceivably, this objection could be overcome through the adoption of special rate structures — not unlike existing standby rates. In general, utilities and state power commissions will need data on the power requirements of wind generator owners before adequate rate schedules can be set.

Selling Excess Power to a Utility

If a wind turbine generator owner requires the connection of his facilities with that of a utility so that he may produce some of his needed power while simultaneously buying the remainder of his needed power, or if he wishes to sell excess power, a proper interconnection between the two generating systems will have to be made. The utility will want to ensure that the connection is safe and will not jeopardize its own facilities or men working on the line. The state commission will oversee this process to its own satisfaction. The utility will require the right to inspect the connection at any subsequent time and make any necessary modifications to ensure safety and proper operation.

Power fed back into the utility's lines will have to be at the correct voltage and be synchronized in frequency and phase. A synchronous inverter is one device available to ensure these conditions. When such an interconnection is made, two meters will probably be used: one to measure the electricity bought from the utility and the other to measure the electricity fed back to the utility's lines. This would enable the utility to charge a particular price for the power it sells and buy the wind turbine generator owner's power at a wholesale rate (if reimbursed at all) to make up for its capital and distribution costs.

With this general description in mind, we will turn to the many legal aspects of this situation. First, it is likely that the WECS owner will have to bear (through the rate structure or otherwise) much of the cost of effecting the interconnection (e.g., the extra meter). Second, it is conceivable (though fairly unlikely) that the utility's general duty to serve all comers may not extend to this situation. Third, a "demand charge" might be applied here (which could involve a higher cents per kilowatt-hour rate *and* an additional charge based on the extra capacity required by the utility) and service might also be interruptible (i.e., capable of being shut off at the utility's option). Fourth, at least some utilities now prohibit a reverse flow of electricity back into the grid when they provide supplementary electricity to a self-supplying customer. Whether this will continue to be the case if the price of conventional fuel increases and wind/solar devices become more numerous is uncertain. Finally, there is a question as to the amount of the credit to be given the WECS owner's bill, assuming that the utility does permit such a sale to it. State utility commissions will probably be required to decide on all of these questions.

WARRANTY, LIABILITY, AND INSURANCE

Liability of the Manufacturers and Installer

Usually one who is injured (financially or physically) by a product can receive money to cover damages on the basis of negligence, warranty, or strict liability. The injured person must show that the product was defective, that the defect caused the injury, and that the defendant being sued is responsible for the defect. The term *defect* has come to mean anything initially wrong with the product that can occur during the process of manufacture and sale. To recover on a claim, an injured person must prove that the defendant should have taken reasonable care to take precautions against creating foreseeable and unreasonable risks of injury to others, and that his not doing so was the cause of the injury (either financial or physical).

In circumstances where there may be a dangerous nature to a product (i.e., blade and tower failure consequences), the seller may have an obligation to give adequate warning of unreasonable dangers of which the seller knows or should know. This obligation to warn the potential buyer extends to all advertising.

An express warranty is a claim, promise, description, or sample made by the seller, which is made part of the bargain with the buyer. The injured person must have knowledge of this claim or promise, and only be injured as a result of reasonable reliance on it. Liability is established when the product is demonstrated to be not as good as was claimed. Potential sources of express warranties include: the name of the product; descriptions of the product found in advertising brochures, catalogs, or packaging; drawings or other pictorial representations accompanying the product; and all representations made by the seller or his agent to the buyer. However, not all claims about a particular product are treated as express warranties giving rise to liability. Mere sales talk and opinion have been distinguished as representations which are not meant to be relied upon by the buyer.

An action for breach of warranty proceeds on a contract theory, as distinguished from the laws governing negligence. As such, it focuses upon the express or implied promises made by the defendant to the injured person and not on the defendant's fault. In such an action a consumer need only prove that the product was defective when sold, did not conform to the defendant's representation about the product, and that he was injured as a result of that defect. The advantage of a warranty action over a negligence claim is the absence of the need to prove that the seller failed to use reasonable care. The rules of warranty have been codified by the Uniform Commercial Code, which has been adopted by all states with the exception of Louisiana.

Whom to sue in a product liability case is a question whose answer depends upon the parties connected with the particular product, as well as plaintiff's evaluation of the economic worth of each potential defendant. In most cases, the defendant will be the manufacturer, distributor, wholesaler, retailer, or other supplier in the direct chain of distribution. Anyone in the chain of distribution who represents a product as his own is subject to the same measure of liability as that of the manufacturer. Liability will depend on whether anyone in the chain of control has the duty to discover defects in the product.

Liabilities of the WECS Owner

Each owner/occupier of land enjoys the privilege of using land for his own benefit. A standard of reasonable care qualifies that privilege by imposing the duty to make a reasonable use of property, which causes no unreasonable harm to others in the vicinity. The duty of reasonable care is affected by the location since the hazards to be anticipated in crowded, commercial areas are not the same as those involved in rural areas. Reasonable care, on the other hand, does not require such precautions as will absolutely prevent injury or render accidents impossible.

Owners of property are under no legal obligation to trespassers other than to do such persons no willful harm. The status of trespasser has been held to include those who enter upon the premises unintentionally, such as persons who wander too far from the highway. The standards applicable in the case of trespassing children, however, are not the same as those for adults. A number of states follow the "attractive nuisance" doctrine, which imposes liability for the creation of conditions that are so alluring to children (despite the danger apparent to those possessing greater discretion) that they are induced to approach and be exposed to the possibility of injury. Liability has been held not to exist in more isolated places where the owner had no reason to anticipate the presence of children.

Insurance

The standard homeowner's insurance package usually covers liability connected with an accessory building with the following conditions attached:

- The installation is not to be used for commercial purposes.
- The structure is not highly susceptible to fire (for example, a woodwork shop or a storage area for flammable materials).

It is not certain if a WECS would be considered to be engaged in commercial activity if excess power is sold to a utility and credit obtained against the cost of power that is bought. Written clarification of this point in such circumstances would assure the owner of necessary coverage. Antennas and masts are not covered against damage from wind, rain, hail, and snow. However, it is possible that a WECS will not be included in the category of antennas and, therefore, will be covered fully against fire and acts of nature that are covered by the policy.

Some insurance underwriters will not want to accept the added risk of a wind turbine and may simply force you to find another insurance company. Those companies that have a sizable business in rural areas with a good wind potential will most likely be prepared to offer coverage for your WECS. The insurance company, of course, will require that the structure conform to all applicable local, state, and federal ordinances and regulations.

Typically, a homeowner's insurance policy provides coverage only for outsiders performing minor amounts of yard work and not jobs that would be covered by workmen's compensation. In this case, help used for the erection of a tower and installation of the wind turbine will not be covered by the homeowner's policy. A contractor will have his own insurance. If other help is used for these tricky and potentially very dangerous jobs, special attention must be paid to insurance coverage.

CHAPTER 8

This Book and Your Wind System – Examples

This chapter provides examples of how wind systems should be selected using the information in this book. Each example focuses on some aspect of the decision-making process, while illustrating the types of calculations you will probably make as a result of asking the questions presented at the beginning of this book. The block diagram of Figure 1–1 will help organize your thinking.

No attempt is made here to reach conclusions with each example; we merely illustrate the steps needed to arrive at the eventual conclusions. Such conclusions will be yours. They will be based on your needs, your wants, your site, and all of the other considerations we have presented. No single hypothetical illustration would do justice to the process of making realistic decisions about your energy source.

EXAMPLE 1

The plan is to replace utility company power, if economical, with a wind-powered electric system for pumping water from deep wells on an alfalfa ranch. The preliminary data gathered by on-site analysis and research are:

- Water depth – 600 feet
- Water requirement – 7 acre-feet of water per acre per growing season
- Growing season – April 1 to October 1
- Energy requirement for pumps already installed – 700 kWh per acre-foot of water

(Note that these data are related directly to the discussion of energy requirements in Chapter 4.)

All preliminary data were supplied by a county agricultural extension agent in the area of the alfalfa ranch. It so happens that this ranch is located very near Palmdale, California – a flat, desert region. Data for Palmdale suitable for a preliminary analysis were presented in Appendix 1. Monthly wind powers are:

April	268 W/m² (watts per square meter)	July	254 W/m²
May	315 W/m²	August	200 W/m²
June	328 W/m²	September	165 W/m²

This data represents two of the planning chart evaluation steps (Figure 1–1): determining the nature of the site wind resource and estimation of energy needs. At this point, the planning chart advises evaluating the social, legal, and environmental impacts of the proposed system, as well as calculating wind turbine sizes and selecting other equipment. Next, refer to Figure 6–1 for system design, where it can be seen that, because of the initial plan, the WECS end use as an electric power generator has already been established. At this point, the ac versus dc question might be pondered, but first return to the environmental concerns.

Suppose, for this example, the following facts are determined by checking at the County Planning Office and the Department of Building and Safety:

- Proposed site not affected by airport air traffic considerations.
- Wind systems (water pumpers) are common in the area and thus have legal precedence.
- Building codes cover tower installations and appear to be easily satisfied by proper engineering of the tower foundation and selection of a tower if the manufacturer can substantiate the tower-design loads by submitting an engineering report to the building department.

A check with the neighbors reveals that no social concern is placed on the proposed installation. This is a farm community and the nearest neighbor is half a mile away. Now, just what does the proposed installation require?

Figure 1–1 instructs us to calculate wind turbine size next (select system components). There are two options in this regard. Existing wind data are in watts per square meter. It could have been in raw windspeed (miles per hour, etc.), which would require the calculations of Chapter 3, but with data in watts per square meter, it is possible to simply calculate the square meters of wind turbine frontal area required to produce the needed watts of power. Here is where the options come in: The wind turbine can be sized to the average of the data, or it can be sized to the minimum wind available.

If the wind turbine is sized to the average month, there will be some months with not enough water and some months with too much water. If, on the other hand, the wind turbine is sized using minimum power available (165 W/m²), then for all the rest of the months there will be a water surplus. Indeed, the wind turbine would be larger than is actually necessary.

Another check with the county agent reveals that water requirements will increase through June, then begin to decrease through September, correlating with the wind curve of Figure 3–3. This indicates that, at least as a preliminary estimate, the average wind power available can be used to size the wind turbine.

The simple average of the wind data can be calculated by adding up the power available each month and dividing by the total number of months—for an average of 255 watts per square meter. At this point two facts should be realized and appropriate action taken:

- Energy requirements were expressed in kWh, not watts or kilowatts
- The wind power numbers are based on 100 percent efficient conversion, not 59.3 percent—the theoretical maximum wind power available.

With regard to the energy source, since power coefficients or efficiency of complete systems (values like 25 to 40 percent) include the 59.3 percent theoretical maximum (re-read Chapter 2, if necessary), it is possible to merely multiply the system efficiency by the average wind power (255 W/m² in this case) to get the actual potential wind which will be harnessed. We only need to estimate a power coefficient for our wind-electric system.

Looking at the table of component efficiencies at the end of Chapter 5, one could guess that the best wind turbine generator will be in the 10 kW-size range. If 30 percent is selected as an estimated efficiency value, actual wind power available would be 76.5 W/m² (0.30 × 255).

To convert the energy requirement from kWh into kW, the number of hours involved must be calculated. Let us assume, for simplicity, that the growing season equals six 30-day months. Then, total hours would be 4320 (24 hours per day × 30 days per month × 6 months).

Further, we need seven acre-feet of water per acre per growing season at 700 kWh per acre-foot. Total energy required would be 4900 kWh per acre per season (700 kWh × 7 acre-feet).

From this, energy demand (4900 kWh) can be converted into an average power demand (kW): 4900 kWh ÷ 4320 hours = 1.13 kW. (Remember, this value is kW per acre.) If it is planned to harvest five acres of alfalfa, 5.65 kW (5 acres × 1.13 kW) total average power capability would be needed from the wind generator. At 76.5 watts per square meter, the wind turbine's size can be calculated as follows:

$$5650 \div 76.5 = 73.9 \text{ square meters frontal area}$$

This could translate to a Darrieus rotor (similar to Figure 5–27) of dimensions 8.5 meters (28 feet) in diameter by 8.5 meters tall, or a propeller-type machine (similar to Figure 5–6) about 9.6 meters (31.9 feet) in diameter. Review the formulas shown in Chapter 5, (Wind Turbine Performance) to calculate the areas and dimensions.

The required size of the wind machine has been calculated based on the average wind power condition. We now need to calculate such a machine's maximum power generation capability in order to move into the next block of Figure 1–1 ("Select system components") which leads to cost analysis.

Checking our wind data, maximum power available will be 98.4 W/m² (328 W/m² × 30 percent efficiency). Then 98.4 W/m² × 73.9 m² = 7272 watts (just over 7 kW). Recall that our first estimate was for a 10 kW (or possibly larger) machine.

By checking the literature from wind turbine manufacturers, we discover a 10-meter-diameter, 8-kW machine manufactured by Brand X to be the nearest to our requirement. It sells for $12,000, including a 90-foot tower.

A preliminary economic evaluation can be made at this point if it is remembered that the raw price of $12,000 does not include installation, batteries, etc. The cost will only increase, but let us see.

A total of 24,500 kWh (4900 kWh per acre × 5 acres) are required per season to grow alfalfa. If the utility presently sells this amount of electricity for 5 cents per kWh, the total seasonal cost would be $1225 (0.05 × 24,500 kWh). We expect to spend $12,000 (or more) on the wind system. We would also like to see a return on our investment of, say, 8 percent per year ($960). On the condition that the wind system will supply all pumping energy needs, it will actually return $1225 by offsetting the cost of power from the utility. This is $265 more than the minimum required $960 return on the investment per year and indicates that the wind system may be economically worthwhile.

The investment return surplus allows leeway for other system costs, including installation, maintenance and storage. We can set up an arbitrary monitor that will serve as a cutoff dollar value of our wind system, above which we will not be willing to spend any more money for the complete system; that is, 8 percent of whose total value will equal $1225. This would be $15,312.

If planning a professional wind system were as simple as this illustration indicates, one would be tempted to take the $15,312 figure to the manufacturers of candidate machines and offer it to them for a completely installed, tested, and guaranteed system. It is not usually that easy, however.

One is also inclined to notice that, if the cost of utility power rises (which it certainly will), the return on already invested money will also rise. This is because the wind turbine is now returning more dollars worth of energy on the same investment. Historically, folks who buy wind systems with no regard for return on invested dollars do not make such an observation. But those individuals and organizations who do expect returns do not observe this fact either. That is, the relationship between bank interest rates (return on investment or cost of money, depending on the situation) and cost of energy is vastly more complex than would allow the simple observation we have offered. All this points to the fact that a business that buys a wind system as a capital item will use an accountant to assess the economic factors involved in the purchase. Our economic analysis of this example system need go no further.

To complete the planning of this system, one would follow the blocks of Figure 1–1, until either the cost of the system rises beyond a limit or the system is completed and its overall economic performance is evaluated as detailed in Appendix 2.

EXAMPLE 2

We very often hear the question "Which should I do — buy a water-pumper windmill or an electric machine? I want to pump my well on wind power!" By now, you probably realize that there is no easy answer. Actually, there are three possible choices: water pumper, wind generator, or none of the above. It makes no sense to consider any of the choices if there isn't enough wind. How much wind do you need? We will discuss that problem shortly. Let's use this

Figure 8-1: A typical farm-type mechanical water pumper.

present example to look at the trade-offs between electric and mechanical pumping; perhaps more closely than we did in Chapter 5.

A valid conclusion can sometimes be drawn from the results of a complete site analysis. Put another way: "Where is the water well?" and "Where is the best windsite?" Answer these two questions and you may have your overall answer. It may be, for example, that the well is located in a sheltered spot, while a great increase in wind energy (an increase, that is, over what is available at the well) is available by locating the wind machine some distance away. Almost any distance, in the case of the classic farm-type water pumper, means that extra mechanical linkages, push rods, and such will have to be installed to transmit power from the wind machine to pump. Here the trade-off starts with mechanical versus electric power transmission. Trade-offs do not stop there, though.

If, on the other hand, the water site and the wind site are one and the same location, the trade-offs may start with cost, aesthetics, or some other consideration. System simplicity will favor the mechanical system. System versatility will favor the electrical system; you can always use the electricity for other things besides pumping water.

Suppose you have a situation where the best water (closest to the surface, hence, easiest to pump) is located at one site where a fair wind speed average has been recorded, whereas a better wind speed average is recorded else-

128

where, at a site where water is known to be somewhat deeper. In such a situation, an intelligent assessment cannot be made without some numbers and a list of objectives and requirements. Here's a sample of such a list.

- Established purpose—water pumping to fill stock ponds for 720 head of range cattle
- Water requirement—estimated daily consumption per head, 10 gallons (estimated from Figure 4–10)
- Total requirement—10 × 720 = 7200 gallons per day
- Site A (good water, less wind): windspeed average = 7 mph, well depth = 60 feet
- Site B (good wind, less water): windspeed average = 11 mph, well depth = 400 feet

From these numbers, we need to pump 7200 gallons per 24-hour day which is 300 gallons per hour, or five gallons per minute (gpm). Our well pipe is two inches in diameter, so, checking Figure 4–8, we find we do not need to consider head loss (friction loss from pushing water through a pipe). By checking Figure 4–9, it can be seen that a five-gallon-per-minute flow at 60-foot depth requires 0.1 hp output from a pump. At 400-foot depth, 0.5 hp output is required. These horsepower requirements are pump *output*. Assuming the pump to be 70 percent efficient, the following pump power *input* requirements can be derived:

$$\text{Site A: } 0.1 \text{ hp} \div 0.7 = 0.14 \text{ hp}$$

$$\text{Site B: } 0.5 \text{ hp} \div 0.7 = 0.71 \text{ hp}$$

Put another way, these new values are the output requirements for our wind machines. These horsepower values can be converted to watts of electric power as follows:

$$\text{Site A: } 0.14 \text{ hp} \times 746 \text{ watts per hp} = 104.4 \text{ watts}$$

$$\text{Site B: } 0.71 \text{ hp} \times 746 \text{ watts per hp} = 529.7 \text{ watts}$$

The wind turbine size required at sites A and B can now be calculated using the equation from Chapter 2:

$$\text{Power} = K \times e \times \text{DRA} \times \text{DRT} \times A \times V^3$$

You may want to review Chapter 2 at this point to refresh your memory concerning this formula. In order to simplify the comparison, we shall assume that our sites are at sea level and the temperature is 60°F. Thus, DRA and DRT both equal 1. The real numbers for your own site can be inserted as indicated in Chapter 2. For this calculation power is measured in horsepower, wind turbine size is measured in square feet of frontal area, and wind speed is measured in miles per hour. For these units we obtain a value for K of 0.00000681.

From the efficiency estimator table near the end of Chapter 5 we can estimate e, the wind turbine efficiency, as 25 percent. For a comparison, just about any assumption will do. To actually calculate system performance, you will need a less casual estimate, but for any small wind system of professional design and manufacture, 0.25 is a safe overall power coefficient estimate.

The resulting calculations are:

Site A: power at the pump = 0.14 hp:
   $0.14 = 0.00000681 \times 0.25 \times A \times 7 \times 7 \times 7$, so
      A = 239.7 square feet, which equals a windwheel of 17.5 feet diameter

Site B: power at the pump = 0.71 hp:
   $0.71 = 0.00000681 \times 0.25 \times A \times 11 \times 11 \times 11$, so
      A = 313.3 square feet, which equals a windwheel of 20 feet diameter

You might be inclined to think that an 18- or 20-foot diameter, high-solidity, water-pump windmill will be a bit on the expensive side. If one could find a used machine of the size needed (in good operating condition at a bargain price), system design would, at least for the trade-off study, end here. But one must also consider the cost evaluation.

There might be a case for setting up a wind machine at Site B, which powers a pump at Site A. This, of course, depends on a parameter not yet introduced into our comparison: distance between sites. If the two sites are not far apart, the analysis could end here. But suppose Site A and Site B are a quarter-mile apart (1320 feet).

Reviewing Chapter 5, we see that candidate methods for transferring power from Site B to Site A include: a mechanical power shaft, hydraulic fluid flow, compressed air, and hydrogen gas (although you would have to pump water to Site B, convert it to hydrogen at Site B to be used at site A to power a pump by fuel cell electricity). Hydraulic fluid pumping and compressed air look very promising since fluid pumps are well matched to windwheels, and aerospace applications are constantly improving hydraulic and pneumatic systems. For this example, let us assume that a manufacturer of a wind turbine generator has offered a system of high-voltage (low line loss) transmission similar to Figure 5–29. If the long wire run is conducted at 400 volts, and we assume a peak power transmission of 500 watts, then our peak amperage is 1.3 amps (500 watts ÷ 400 volts).

Looking at Figure A2–4 and the formula for line loss and wire size, we decide that a two-percent line loss (8 volts) is acceptable. To calculate wire size, the following calculation must be made:

Circular area size = 35×1.3 amps $\times 1320$ ft.
(aluminum wire)

$\div$ 8 volts = 7507 circular mills

Looking at the chart in Appendix 2 (Wiring section), we see that this falls between number 12 and number 10 size. This, we decide, is acceptable. While it is beyond the scope of data in this book, we can expect another 5 percent loss from the transformers. To be safe, let us assume a total of 10 percent line loss. Then, to calculate total average power required:

104.4 watts $\times 1.10 = 114.8$ watts

Now we calculate wind turbine size as before but with different units in the equation:

$$114.8 \text{ watts} = 0.00508 \times 0.25 \times A \times 11 \times 11 \times 11$$

so:

$$A = 67.9 \text{ ft.}^2$$

This equals a wind turbine just under 10 feet in diameter. A machine with a 10-foot diameter rotor, in this example, mounted on Site B can replace a 20-foot machine at Site B or an 18-foot machine at Site A (if the water is being pumped at Site A).

If you perform a trade-off analysis like this, you will follow the above conclusions with a more detailed study, which will determine whether the added complexity and cost of the long electrical transmission lines is offset by the reduced complexity and cost of the remote-site wind machine. This would be the remaining question, and it requires a survey of specific, on-the-market equipment. Elsewhere in this book we provide you with the addresses of organizations that keep up-to-date lists of sources of equipment you will need.

Glossary

ac	alternating electric current
airfoil	a curved surface designed to create lift as air flows over its surface
amp-hours	see *amps*, calculated by multiplying current flow by number of hours it flows
amperes or amps	a measure of electric current flow
anemometer	an instrument for measuring wind speed
asynchronous generator	an electric generator designed to produce an alternating current that matches an existing power source (e.g., utility mains) so the two sources can be combined to power one load (e.g., your home). The generator does not have to turn at a precise rpm to remain at correct frequency or phase; see also *synchronous generator.*
current	flow of electricity through wires
cut-in speed	wind speed at which wind turbine begins to produce power
cut-out speed	wind speed at which wind turbine is shut down to prevent high wind damage. Also furling speed.
dc	direct electric current; does not alternate direction of electric flow as does ac
diode	see also *rectify*, an electric device that changes ac to dc
drag	a force that "slows down" the motion of wind turbine blades, or actually causes motion and power to be produced by drag-type wind machines
efficiency (e)	a number arrived at by dividing the power output of a device by the power input to that device (usually the larger of the two numbers); usually expressed as a percentage value; see also *power coefficient;*
energy	a measure of the amount of work that can be, or has been done; expressed in kilowatt-hours (kWh) or horsepower-hours (hphr)
energy density	a ratio of energy per pound; a rating usually used to compare different batteries
energy rose	see also *wind rose*, a diagram that presents wind energy measurements from a site analysis in relation to the direction from which the wind occurs at the site
fantail	a propeller-type wind turbine mounted sideways on a larger wind machine (horizontal-axis type) to keep that machine aimed into the wind
furling speed	the wind speed at which the wind machine must be shut down to prevent high wind damage. Also cut-out speed.

gear ratio	a ratio of speeds (rpm) between the rotor power shaft and the pump, generator, or other device power shaft; applies both to speed-increasing and speed-decreasing transmissions
gin pole	a pipe, board, or tower used to improve leverage while raising a tower
head	a measure of height a pump must lift water
head loss	a measure of friction loss caused from water flow through pipes
horsepower (hp)	a measure of power; 550 pounds raised one foot per one second
horsepower-hour (hphr)	a measure of energy, see also *energy*
inverter	a device that converts dc to ac; generates its own frequency and voltage references; see also *synchronous inverter*
kilowatt (kW)	a measure of power; one horsepower equals 776 watts, or 0.776 kilowatts
kilowatt-hour (kWh)	a measure of electric energy, (1000 watt-hours), see also *kilowatt, horsepower,* and *horsepower-hour*
lift	the force which "pulls" a wind turbine blade along, as opposed to drag
megawatt	one million watts
meteorological station	location where the weather is recorded
panemone	a name for drag-type vertical axis wind machines; coming from *pan* (all directions) and *anemone* (wind), it could describe Darrieus-type machines also, but is not generally used except for drag machines
power	the rate work is performed — mechanical power is force times velocity (see *horsepower*); electric power is volts times amps
power coefficient	ratio of power output to power input; often referred to as efficiency
rated power	the power output (watts or horsepower) of a wind machine; can be its maximum power, or a power output at some wind speed less than the maximum speed before governing controls reduce the power
rated speed	wind speed at which rated power occurs; can be speed at which a governor takes over, or can be a wind speed lower than this; an industry standard for rated speed does not exist at this time
rectify	convert ac to dc, see also *diodes*
resistor	an electric device that "resists" electric current flow, used to control current (e.g., field-current in a generator)
return time	time before the wind returns to a higher, specified value, such as the cut-in speed of a windmill
rotor	the power producing structure of a wind turbine (e.g., the blades)
rotor efficiency	the efficiency of the rotor only; does not include transmissions, pumps, generators, or line or head loss
rotor power coefficient	same as rotor efficiency
run of the wind	the distance the wind travels during a specific time period; this usually refers to the dial reading from a wind anemometer
shelter belt	a tree row planted in windy country to shelter crops and soil
sine wave	the type of ac generated by utility companies, rotary inverters, sophisticated solid-state inverters, and ac generators
solidity	ratio of rotor blade surface area to frontal (or swept) area of the entire windwheel

square wave	type of ac output from low-cost, solid-state inverters; usable for many appliances, but may affect stereos and TV sets
synchronous generator	an ac generator that operates together with ac power source (similar to *asynchronous generator*) must turn at a precise rpm to hold frequency and phase relationship to the ac source
synchronous inverter	also called "line commutated inverter" inverts dc to ac (see *inverter*) but must have another ac source (e.g., utility mains, or ac gas generator) for voltage and frequency reference; ac is created synchronously, that is, in phase and at same frequency as outside ac source
torque	a measure of force from windwheel causing rotary motion of power shaft
turbulence	rapid wind speed fluctuations; gusts are maximum values of wind turbulence; randomness in the wind
voltage	the electrical pressure which causes current flow (amps)
watt	unit of electric power, see also *horsepower*
watt-hour	unit of electric energy, see also *kilowatt-hour*
watt per square meter	a measure of the energy in the wind passing through a square meter of area
WECS	wind energy conversion system
windmill	archaic term for wind system; still used to refer to high-solidity rotor water pumpers and older mechanical output machines
wind power	power in the wind, part of which can be extracted by a wind turbine; see *power*
wind power profile	how the wind power changes with height above the surface of the ground or water; typical plots of wind power profiles for flat terrain with different types of plant and tree cover are presented in Figure 3–14; the wind power profile is proportional to the cube of the wind speed profile (see *wind speed profile*)
wind rose	a plot showing the average (usually a monthly or yearly average) wind speed from each direction (usually 16 directions are used) and percentage time the wind blows from each direction
wind speed profile	how the wind speed changes with height above the surface of the ground or water; see Appendix 4.
wind turbine, wind system, or wind machine	accepted modern terms for devices which extract power from the wind; can refer to devices which produce mechanical or electrical output
wind turbine generator	a wind system that produces electrical power; abbreviated WTG
windwheel	same as *rotor*
work	force lined up with the direction of movement times the distance moved; for example, by lifting a 1-pound weight up 1 foot, 1 foot-pound of work is performed. A 2-pound weight lifted up 3 feet requires $2 \times 3 = 6$ foot-pounds of work.

Additional References on Wind Energy

Periodicals

A WEA Newsletter, American Wind Energy Association, c/o Secretary, AWEA, 1609 Connecticut Ave., Washington, D.C. 20009, $25 a year (includes membership in the association).

Alternative Sources of Energy, ASE, Route 2, P.O. Box 90-A, Milaca, MN 56353, $6 a year.

RAIN: Journal of Appropriate Technology, RAIN, 2270 N.W. Irving, Portland, OR 97210, $10 a year.

Wind Power Digest, Jester Press, 109 East Lexington Ave., Elkhart, IN 46514, $8 a year.

Windustries, Great Plains Windustries, Inc., P.O. Box 126, Lawrence, KS 66044, $10 a year, $15 a year for institutions.

Books

Clews, Henry. *Electric Power From The Wind*. Solar Wind Co., P.O. Box 7, East Holden, ME 04429, $2. A brief overview of the requirements for a complete wind system.

Eldridge, Frank. *Wind Machines*. For sale by the Superintendent of Documents, U.S. Government Printing Office, Washington, D.C. 20402 (stock no. 038-000-00272-4), $4.25; well-illustrated historical and technical document on the background of wind machines.

Park, Jack. *Simplified Wind Power Systems for Experimenters*. Helion, Inc., P.O. Box 445, Brownsville, CA 95919, $6. Written for designers and technically oriented readers.

The Power Company-Midwest, Inc. *The Windcyclopedia*. Windcyclopedia, P.O. Box 221, Gennessee Depot, Wisconsin, 53127, $7.95; a comprehensive collection of wind energy information sources and systems manufacturers.

Wegley, H., Orgill, M. and Drake, R. *A Siting Handbook for Small Wind Energy Conversion Systems*. Battelle Pacific Northwest Laboratories; available from the National Technical Information Service, 5285 Port Royal Road, Springfield, VA 22161, $7.

135

APPENDIX 1

Wind Power in the United States and Southern Canada

The following table lists the wind power at 750 stations in the United States and Southern Canada. The data in the table have been extracted from the report *Wind Power Climatology in the United States* by Jack Reed of Sandia Laboratories, Albuquerque, New Mexico (June 1975). This report (SAND 74-0348) can be ordered from the National Technical Information Service, U.S. Department of Commerce, 5285 Port Royal Road, Springfield, VA 22151. A printed copy costs $7.60 and a microfiche copy costs $2.25 (see if your local library has a microfiche reader). Besides the data in this table, the report contains average monthly results for each station of the percentage of time the velocity was in each of about eight speed ranges, i.e.:

January Wind Velocity (mph)

0–3	4–7	8–12	13–18	19–24	25–31	32–38	39–46
27.6%	26.1	23.5	18.6	3.5	0.6	0.2	0

The data have not been corrected for varying heights of the wind anemometer. Also, possible distortions in the wind pattern by natural terrain features, trees, and buildings are not accounted for. Because of this, no particular set of these data can be blindly accepted as representative of a particular region (see discussion in Chapter 3).

The stations are listed by region within each state. The states are listed first, alphabetically, then the Southern Canadian provinces. Listed on each line are the following:

1. State – U.S. Postal abbreviation (obvious abbreviations of Canadian provinces).
2. Location – the most common abbreviations are APT (airport), AFB (Air Force Base), AFS (Air Field Station), IAP (International Airport), IS (island), NAF (Naval Air Field), PT (point), WBO (Weather Bureau Office).
3. International station number.
4. Latitude in degrees and minutes (3439 = 34° 39′ N).
5. Longitude in degrees and minutes (8646 = 86° 46′ W).
6. Average wind speed in knots (multiply by 1.15 to convert to mph).
7. Twelve average monthly wind power values in watts per square meter (multiply by 0.0929 to convert to watts per square foot).
8. Average of the previous 12 monthly power values.

136

Monthly Average Wind Power in the United States and Southern Canada*

State	Location	Sta. No.	Lat	Long	Ave. Speed knots	J	F	M	A	M	J	J	A	S	O	N	D	Ave.
AL	Huntsville	3856	3439	8646	6.6	80	109	118	87	48	37	29	31	56	50	78	87	60
AL	Foley	93826	3358	8605	8.0	133	182	153	174	142	106	73	66	109	96	116	115	122
AL	Gadsden	75258	3358	8605	5.8	84	104	114	99	43	34	25	21	40	45	63	63	61
AL	Birmingham APT	13876	3334	8645	7.3	127	157	156	137	80	64	49	44	68	68	108	106	97
AL	Tuscaloosa, Vn D Graf APT	93806	3314	8737	5.1	79	79	93	69	33	21	15	21	28	36	55	69	49
AL	Selma, Craig AFB	13850	3221	8659	5.7	74	86	91	68	42	34	29	26	37	31	48	54	51
AL	Montgomery	13895	3218	8624	6.1	74	90	85	68	39	35	34	26	39	36	51	62	53
AL	Montgomery, Maxwell AFB	13821	3223	8621	4.8	58	67	69	51	28	25	20	19	27	26	39	45	39
AL	Ft. Rucker, Cairns AAF	3850	3116	8543	4.7	40	50	55	41	23	18	12	12	19	19	29	35	30
AL	Evergreen	13885	3125	8702	5.3	66	69	78	57	29	18	17	17	21	24	38	51	40
AL	Mobile, Brookley AFB	13838	3038	8804	7.3	105	104	128	119	94	58	43	41	69	51	72	89	80
AK	Annette IS	25308	5502	13134	9.5	320	264	216	199	110	97	71	77	128	297	324	319	199
AK	Ketchikan	952	5521	13139	5.8	59	53	42	52	47	37	38	44	43	67	75	75	57
AK	Craig	25317	5529	13309	7.9	185	159	167	132	82	95	71	55	113	186	174	165	128
AK	Petersburg	960	5649	13257	3.7	26	40	37	41	32	23	22	22	24	29	22	21	33
AK	Sitka	961	5703	13520	3.5	109	26	34	42	27	23	22	14	34	44	46	93	37
AK	Juneau APT	25309	5822	13435	7.5	119	134	123	127	95	70	60	67	108	170	157	159	115
AK	Haines	955	5914	13526	8.0	218	202	203	148	74	61	94	54	72	160	238	159	146
AK	Yakutat APT	25339	5931	13940	7.0	177	144	114	100	90	71	56	64	96	181	183	169	114
AK	Middleton IS AFS	25403	5927	14619	11.9	625	597	468	355	238	141	96	134	243	519	588	608	376
AK	Cordova, Mile 13 APT	26410	6430	14530	4.4	46	48	42	41	37	23	18	17	32	53	47	48	36
AK	Valdez	26442	6107	14616	4.3	72	28	75	41	36	16	13	7	7	45	100	172	53
AK	Anchorage, IAP	26451	6110	15001	5.9	61	95	48	61	108	76	61	52	46	38	38	50	61
AK	Anchorage, Merrill Fld	26409	6113	14950	4.9	57	66	29	27	41	40	23	22	30	30	59	23	37
AK	Anchorage, Elmendorff AFB	26401	6115	14948	4.4	46	60	50	41	40	34	24	22	26	30	46	33	36
AK	Kenai APT	26523	6034	15115	6.6	96	109	94	66	61	63	56	54	53	83	85	80	74
AK	Northway APT	26412	6257	14156	3.9	16	21	30	44	40	42	33	32	27	22	18	16	28
AK	Gulkana	26425	6209	14527	5.8	45	88	85	105	111	98	83	100	95	76	48	40	81
AK	Big Delta	26415	6400	14544	8.2	447	322	239	147	148	85	68	102	163	209	300	333	215
AK	Fairbanks IAP	26411	6449	14752	4.3	10	16	25	37	50	44	33	29	28	22	13	10	27
AK	Fairbanks, Ladd AFB	26403	6451	14735	3.5	10	17	24	28	38	35	23	29	23	24	12	9	23
AK	Ft. Yukon APT	26413	6634	14516	6.7	30	41	64	81	91	84	86	81	74	52	31	31	64
AK	Nenana APT	26435	6433	14905	5.1	68	44	48	45	46	34	27	26	33	42	45	44	42
AK	Manley Hot Springs	567	6500	15039	4.8	76	54	84	109	93	89	53	42	63	104	62	52	62
AK	Tanana	976	6510	15206	6.6	22	85	89	83	56	53	47	29	50	64	56	80	73
AK	Ruby	508	6444	15526	6.5	58	133	119	84	40	51	46	42	64	76	119	54	78
AK	Galena APT	26501	6444	15656	5.4	56	69	66	77	51	53	48	61	61	59	59	49	59
AK	Kaltag	458	6420	15845	4.7	46	103	26	81	31	33	30	28	37	56	40	51	56
AK	Unalakleet APT	26627	6353	16048	10.5	520	502	336	191	112	96	116	146	175	234	395	376	265
AK	Moses Point APT	26620	6412	16203	10.6	329	363	275	279	149	129	181	233	217	222	246	263	241
AK	Golovin	502	6433	16302	9.6	188	229	246	250	142	117	178	264	271	258	369	256	236
AK	Nome APT	26617	6430	16526	9.7	328	308	228	225	153	119	117	162	189	230	263	238	217
AK	Northeast Cape AFS	26632	6319	16858	11.0	468	263	246	347	239	137	218	240	288	462	632	387	328
AK	Tin City AFS	26634	6534	16755	15.0	763	919	811	658	427	271	260	334	352	522	722	728	549
AK	Kotzebue	26616	6652	16238	11.2	455	418	310	294	161	187	212	234	228	270	397	366	291

*See previous page for an explanation of column headings.

Monthly Average Wind Power in the United States and Southern Canada (Continued)

State	Location	Sta. No.	Lat	Long	Ave. Speed knots	J	F	M	A	M	J	J	A	S	O	N	D	Ave.
AK	Cape Lisburne AFS	26631	6853	16608	10.5	432	268	335	266	227	210	303	216	266	432	444	333	314
AK	Indian Mountain AFS	26535	6600	15342	5.4	115	113	88	58	57	37	30	36	52	86	95	104	70
AK	Bettles APT	26533	6655	15131	6.3	28	44	57	62	66	62	44	38	43	43	43	47	48
AK	Wiseman	979	6726	15013	3.1	28	26	16	15	24	15	23	14	12	12	26	16	23
AK	Umiat	26537	6922	15208	6.0	113	121	43	77	80	93	62	53	57	51	121	78	76
AK	Point Barrow	27502	7118	15647	10.5	215	194	162	167	169	143	145	208	211	258	286	183	192
AK	Barier IS	26401	7008	14338	11.3	512	468	379	279	216	145	123	208	287	470	486	425	341
AK	Sparrevohn AFS	26534	6106	15534	4.7	69	73	108	76	47	35	36	41	54	63	74	82	63
AK	McGrath	26510	6258	15537	4.2	13	27	29	37	39	35	34	35	32	24	15	12	27
AK	Tataline AFS	26536	6253	15557	4.4	25	37	36	37	38	27	27	29	33	35	24	21	31
AK	Flat	16	6229	15805	8.1	206	266	205	150	116	100	81	108	143	168	185	184	172
AK	Aniak	26516	6135	15932	5.6	51	59	63	59	49	37	27	34	41	47	47	42	46
AK	Bethel APT	26615	6047	16148	9.8	229	258	224	166	125	108	110	137	140	158	185	211	171
AK	Cape Romanzof AFS	26633	6147	16602	11.7	692	699	493	476	246	124	110	154	234	305	520	654	380
AK	Cape Newenham AFS	25623	5839	16204	9.8	400	371	330	288	168	119	101	142	165	212	300	315	241
AK	Kodiak FWC	25501	5744	15231	8.8	328	271	258	198	124	87	52	77	120	210	294	329	189
AK	King Salmon APT	25503	5841	15639	9.2	250	260	235	180	182	138	92	139	156	180	230	206	191
AK	Port Heiden APT	25508	5657	15837	12.9	576	564	493	361	289	273	225	381	466	451	439	565	429
AK	Port Mollor	25625	5600	16031	8.8	158	168	171	195	135	81	108	144	164	222	260	219	172
AK	Cold Bay APT	25624	5512	16243	14.6	736	731	699	580	506	465	428	507	462	606	652	631	573
AK	Dutch Harbor NS	25611	5353	16632	9.6	355	376	295	223	135	125	69	105	169	390	419	266	233
AK	Driftwood Bay	25515	5358	16651	8.0	204	203	154	148	115	72	88	77	71	120	161	182	131
AK	Umnak IS, Cape AFB	25602	5323	16754	13.5	651	688	577	514	454	251	163	249	466	603	606	723	497
AK	Nikolski	25626	5255	16847	14.0	538	560	532	566	437	321	239	283	361	634	732	662	482
AK	Adak	25704	5153	17638	12.2	426	467	528	453	366	223	218	258	331	502	481	525	404
AK	Amchitka IS	45702	5123	17915	18.0	1764	1517	1418	1062	653	448	405	457	740	1053	1165	1569	1025
AK	Attu IS	45709	5250	17311	11.2	553	582	508	403	235	162	135	129	360	366	414	554	368
AK	Shemya APT	45715	5243	17406	15.7	887	932	878	641	483	266	235	285	432	301	977	870	633
AK	St. Paul IS	25713	5707	17016	15.0	758	867	684	518	355	207	175	282	399	693	691	791	547
AZ	Grand Canyon	378	3557	11209	6.2	38	43	49	71	66	55	35	31	57	58	44	28	49
AZ	Winslow APT	23194	3501	11044	7.3	104	104	232	169	161	141	93	77	63	73	63	78	111
AZ	Flagstaff, Pulliam APT	3103	3508	11140	6.4	71	70	96	95	93	86	40	33	52	56	69	69	69
AZ	Maine	178	3509	11157	8.9	132	186	218	253	240	224	111	68	116	178	139	158	151
AZ	Ashfork	171	3514	11233	7.5	111	116	154	201	142	126	82	68	86	100	103	82	114
AK	Kingman	381	3516	11357	8.9	126	156	172	203	153	166	126	99	102	124	115	107	138
AZ	Prescott	23184	3439	11226	7.5	54	95	117	144	138	124	75	56	67	57	59	44	85
AZ	Yuma APT	23195	3240	11436	6.8	55	62	68	77	71	69	93	77	45	40	55	51	62
AZ	Phoenix	23183	3326	11201	4.8	16	28	34	39	37	35	43	31	28	24	22	17	29
AZ	Phoenix, Luke AFB	23111	3332	11223	4.6	21	31	41	52	49	43	49	39	25	21	20	18	34
AZ	Chandler, William AFB	23104	3318	11140	4.1	17	21	28	35	34	33	41	33	28	22	18	16	26
AZ	Tucson APT	23160	3207	11056	7.3	71	59	69	90	87	73	75	54	62	78	82	72	74
AZ	Tucson	23160	3207	11056	7.1	71	59	69	90	87	73	75	54	62	78	82	72	74
AZ	Tucson, Davis-Monthan AFB	23109	3210	11053	5.7	48	48	57	63	56	60	51	35	42	40	43	45	49
AZ	Ft. Hauchuca	3124	3134	11020	5.7	49	58	84	96	80	66	39	27	30	30	34	41	53
AZ	Douglas	93026	3128	10937	6.4	84	94	166	143	128	86	61	46	47	63	62	75	87

State	Location	Sta. No.	Lat	Long	Ave. Speed knots	J	F	M	A	M	J	J	A	S	O	N	D	Ave.
AR	Walnut Ridge APT	93991	3608	9056	6.0	93	81	103	104	58	44	27	22	28	43	64	75	62
AR	Blytheville AFB	13814	3558	8957	6.4	85	106	108	111	66	41	26	25	36	39	67	71	65
AR	Ft. Smith APT	13964	3520	9422	7.4	76	86	116	104	81	62	51	45	50	55	66	75	73
AR	Little Rock	13963	3444	9214	7.6	82	91	105	96	70	58	46	46	48	50	73	71	70
AR	Jacksonville, Ltl. Rk. AFB	3930	3455	9209	5.8	61	67	85	70	47	34	28	24	28	29	45	49	48
AR	Pine Bluff, Grider Fld	93988	3410	9156	6.5	102	89	102	87	51	39	31	29	35	45	71	81	64
AR	Texarkana, Webb Fld	13977	3327	9400	7.7	92	108	128	115	77	69	48	49	62	61	74	87	80
CA	Needles APT	23179	3446	11437	6.7	108	125	128	112	108	98	67	67	58	78	113	124	97
CA	El Centro NAAS	23199	3249	11541	7.7	98	126	171	208	225	189	80	73	79	86	98	76	127
CA	Thermal	3104	3338	11610	9.1	66	79	103	149	191	153	125	114	119	92	76	63	111
CA	Imperial Bch., Ream Fld	93115	3234	11707	5.9	48	51	54	54	52	45	35	32	30	31	43	42	43
CA	San Diego, North IS	93112	3243	11712	5.3	32	41	56	56	49	41	33	32	35	31	30	32	39
CA	San Diego	23188	3244	11710	5.4	30	33	40	47	47	40	30	29	27	25	22	21	31
CA	Miramar NAS	93107	3252	11707	4.4	23	24	28	30	26	19	16	17	18	19	20	24	22
CA	San Clemente IS NAS	93117	3301	11835	6.3	53	72	89	97	67	48	33	32	33	32	54	69	55
CA	San Nicholas IS	93116	3315	11948	9.9	152	199	295	306	348	244	161	166	164	140	180	159	209
CA	Camp Pendleton	3154	3313	11724	5.2	30	35	45	61	53	43	43	44	36	24	28	29	38
CA	Oceanside	189	3318	11721	8.0	129	122	108	82	67	59	49	49	64	66	96	116	87
CA	Laguna Beach	195	3332	11747	5.0	35	38	44	37	30	30	27	26	25	25	22	32	34
CA	El Toro MCAS	93101	3340	11744	4.8	45	38	33	30	26	22	19	19	19	23	36	43	28
CA	Santa Ana MCAF	93114	3342	11750	4.6	43	43	47	46	37	31	30	26	25	26	36	43	37
CA	Los Alamitos NAS	93106	3348	11807	4.8	36	39	47	44	41	32	28	25	22	23	36	37	34
CA	Long Beach APT	23129	3349	11809	4.9	27	40	45	48	43	35	34	32	31	27	29	26	35
CA	Los Angeles IAP	23174	3356	11824	5.9	40	57	69	70	63	49	43	44	39	36	38	35	48
CA	Ontario	93180	3404	11737	7.7	36	117	109	118	148	124	135	135	92	71	46	127	103
CA	Riverside, March AFB	23119	3353	11715	4.4	35	43	40	44	49	51	52	49	37	28	29	32	41
CA	San Bernardino, Norton AFB	23122	3406	11715	3.5	43	43	33	27	25	22	22	20	19	17	28	29	28
CA	Victorville, George AFB	23131	3435	11723	7.7	99	134	170	183	163	135	87	85	74	70	87	90	118
CA	Daggett	23161	3452	11647	9.6	94	173	315	290	355	236	177	159	145	121	107	74	187
CA	China Lake, Inyokern NAF	93104	3541	11741	7.1	121	156	238	249	225	186	124	126	113	124	103	93	155
CA	Muroc, Edwards AFB	23114	3455	11754	7.9	90	118	187	206	236	230	155	131	99	87	82	83	141
CA	Palmdale	81	3438	11806	10.2	163	205	226	267	315	328	254	200	165	158	130	109	226
CA	Palmdale APT	23182	3438	11805	8.8	121	146	233	234	234	229	173	141	107	104	113	132	163
CA	Saugus	83	3423	11832	6.3	105	128	88	96	96	108	101	88	67	76	105	90	89
CA	Van Nuys	23130	3413	11830	4.6	105	82	66	50	43	21	22	19	18	22	90	69	49
CA	Oxnard AFB	23136	3413	11905	4.4	63	56	49	46	43	26	20	19	19	31	50	76	41
CA	Point Mugu NAS	93111	3407	11907	5.6	100	79	71	78	51	33	28	26	28	35	76	82	55
CA	Santa Maria	23273	3454	12027	6.5	75	80	114	94	93	93	63	57	56	66	79	91	82
CA	Vandenberg, Cooke AFB	93214	3444	12034	6.1	62	67	99	97	115	67	34	33	41	51	58	58	65
CA	Pt. Arguello	93215	3440	12035	7.2	72	105	138	135	133	79	58	54	51	74	76	66	85
CA	San Louis Obispo	93206	3514	12039	6.9	60	69	134	127	146	173	105	120	131	129	89	73	115
CA	Estero	395	3526	12052	4.3	83	66	69	76	60	50	22	31	42	47	44	77	53
CA	Paso Robles, Sn Ls Obispo	93209	3540	12038	5.5	34	39	57	76	105	127	106	83	59	42	32	30	64
CA	Jolon	93218	3600	12144	2.8	9	6	10	6	11	11	8	8	6	4	6	6	7
CA	Monterey NAF	23245	3635	12152	5.0	30	33	45	48	51	45	35	32	23	21	20	30	35

Monthly Average Wind Power in the United States and Southern Canada (Continued)

State	Location	Sta. No.	Lat	Long	Ave. Speed knots	J	F	M	A	M	J	J	A	S	O	N	D	Ave.
CA	Ft. Ord, Fritzsche AAF	93217	3641	12146	5.7	30	31	46	61	67	63	66	59	41	34	25	24	47
CA	Taft, Gardner Fld	23126	3507	11918	4.4	20	18	20	29	45	46	31	22	18	16	18	17	26
CA	Bakersfield, Meadows Fld	23155	3525	11903	5.4	27	33	46	55	69	65	47	43	34	25	24	28	41
CA	Bakersfield, Minter Fld	23102	3530	11911	5.0	26	31	38	49	61	73	38	25	22	21	19	25	34
CA	Lemoore NAS	23110	3620	11957	4.8	21	30	38	40	45	47	35	29	25	27	18	19	30
CA	Fresno, Hammer Fld	93193	3646	11943	5.5	24	28	42	48	60	62	42	33	25	23	17	20	35
CA	Bishop APT	23157	3722	11822	7.5	74	106	161	145	129	100	80	81	85	101	88	80	103
CA	Merced, Castle AFB	23203	3722	12034	6.0	56	66	72	74	69	78	59	52	44	44	34	42	59
CA	Livermore	196	3742	12147	7.9	109	108	115	124	158	180	173	143	107	85	64	71	122
CA	San Jose APT	23293	3722	12155	6.4	51	47	61	61	86	84	52	43	46	34	45	47	54
CA	Sunnyvale, Moffett Fld	23244	3725	12204	5.4	47	50	54	59	65	73	62	54	41	35	32	56	51
CA	San Francisco IAP	23234	3737	12223	9.5	96	129	183	228	268	280	236	211	171	141	80	91	176
CA	Farallon IS	495	3740	12300	9.6	61	406	287	193	188	208	100	91	83	106	204	275	212
CA	Alameda FWC	23239	3748	12210	7.4	92	94	122	125	129	124	99	87	65	65	69	81	93
CA	Oakland	23230	3744	12212	6.8	52	75	77	92	101	98	74	69	57	50	40	51	71
CA	San Rafael, Hamilton AFB	23211	3804	12231	4.8	51	52	54	52	50	50	39	39	30	34	32	51	45
CA	Fairfield, Travis AFB	23202	3816	12156	10.7	114	153	176	232	347	488	577	481	332	182	106	91	270
CA	Point Arena	499	3855	12342	13.0	401	398	361	488	500	614	388	513	321	368	320	467	421
CA	Sacramento	23232	3831	12130	7.8	145	145	126	118	116	128	92	83	64	70	61	123	95
CA	Sacramento, Mather AFB	23206	3834	12110	6.0	117	108	89	69	63	69	56	45	38	46	59	88	72
CA	Sacramento, McClellan AFB	23208	3840	12124	6.5	107	102	98	79	83	90	62	56	49	67	75	84	79
CA	Auburn	190	3857	12104	8.4	106	148	109	76	77	64	65	65	64	57	69	67	83
CA	Blue Canyon APT	23225	3917	12042	8.4	237	212	168	106	92	75	57	64	65	110	130	188	122
CA	Donner Summit	23226	3920	12022	12.1	1100	619	729	269	266	226	173	168	154	439	579	645	463
CA	Beale AFB	93216	3908	12126	5.1	75	59	64	56	49	52	31	29	34	39	43	62	50
CA	Williams	498	3906	12209	8.2	163	172	179	112	126	120	78	64	78	105	112	116	111
CA	Ft. Bragg	590	3927	12349	5.9	75	88	82	96	46	44	25	26	25	33	50	52	51
CA	Eureka, Arkata APT	24283	4059	12406	6.0	93	93	109	102	115	87	56	42	39	50	61	75	75
CA	Mt. Shasta	595	4116	12216	11.9	456	535	349	309	343	297	177	163	182	214	295	262	309
CA	Redding	592	4034	12224	7.9	71	86	94	81	88	89	69	62	68	68	72	70	74
CA	Montague	197	4144	12231	5.8	65	130	120	122	130	131	123	100	76	75	71	57	98
CA	Montague, Siskiyou Co APT	24259	4146	12228	5.3	106	108	123	115	78	63	59	50	45	64	82	89	80
CO	La Junta	23067	3803	10331	8.3	115	136	222	204	168	164	94	84	85	78	139	115	134
CO	Alamosa APT	23061	3727	10552	7.4	92	110	195	254	214	167	84	70	85	91	77	74	127
CO	Pueblo, Memorial APT	93058	3817	10431	7.7	101	122	180	231	168	129	105	84	82	81	93	104	121
CO	Colo Springs, Peterson Fld	93037	3849	10443	9.0	142	163	217	212	189	163	99	86	105	105	138	128	142
CO	Ft. Carson, Butts AAF	94015	3841	10446	7.3	85	93	145	218	127	131	63	71	68	112	74	87	107
CO	Denver	23062	3945	10452	8.8	117	139	182	183	132	126	94	83	85	88	118	136	126
CO	Denver, Lowry AFB	23012	3943	10454	8.1	115	94	131	163	112	100	95	87	102	88	126	121	109
CO	Aurura Co, Buckley Fld	23036	3942	10445	6.7	60	60	79	121	79	67	54	52	51	51	57	59	66
CO	Akron, Washington Co APT	24015	4010	10313	11.7	216	313	383	359	276	239	226	184	243	212	252	280	242
CO	Rifle Co, Garfield Co APT	23069	3932	10744	4.1	17	32	37	69	51	39	26	23	31	25	23	15	31
CO	Craig	24046	4031	10733	7.7	57	63	70	97	80	58	50	52	54	61	56	51	62
CT	Hartford, Bradley Fld	14740	4156	7241	7.7	115	127	142	129	96	75	54	53	61	74	93	100	93
CT	New Haven, Tweed APT	14758	4116	7253	8.7	117	122	142	120	83	65	52	60	78	89	114	106	98

State	Location	Sta. No.	Lat	Long	Ave. Speed knots	J	F	M	A	M	J	J	A	S	O	N	D	Ave.
CT	Bridgeport APT	94702	4110	7308	10.4	244	274	256	219	158	114	96	101	139	192	214	251	186
DE	Dover AFB	13707	3908	7528	7.7	135	152	148	125	85	69	49	49	73	78	99	104	96
DE	Delaware Breakwater	404	3848	7506	12.7	449	570	477	430	283	196	163	190	270	403	391	410	343
DE	Wilmington, New Castle APT	13781	3940	7536	8.1	127	149	175	147	105	85	66	59	61	84	118	126	109
DC	Washington, Andrews AFB	13705	3848	7653	7.2	130	156	161	126	77	51	39	36	45	62	101	109	90
DC	Washington, Bolling AFB	13710	3850	7701	7.5	125	173	171	140	84	58	45	40	51	72	119	112	101
DC	Washington National	13743	3851	7702	8.6	142	151	163	134	95	82	62	44	67	85	103	107	105
DC	Washington, Dulles IAP	93738	3857	7727	6.7	104	115	118	111	66	42	37	41	40	42	66	78	68
FL	Key West NAS	12850	2435	8147	9.5	158	172	172	176	122	98	78	71	133	133	139	147	131
FL	Homestead AFB	12826	2529	8023	6.4	61	74	90	89	72	51	31	35	66	59	60	56	60
FL	Miami	12839	2548	8016	7.8	87	98	111	116	80	59	58	54	90	88	78	79	80
FL	Boca Raton	12803	2622	8006	8.2	80	108	125	135	109	72	51	55	109	140	108	106	99
FL	West Palm Beach	12865	2643	8003	8.3	123	129	151	145	106	79	70	67	80	105	126	102	108
FL	Ft. Myers	12835	2635	8152	7.0	93	111	153	156	104	79	58	70	99	96	90	101	101
FL	Ft. Myers, Hendricks Fld	12802	2638	8142	7.1	59	68	98	91	74	51	38	47	76	85	58	60	69
FL	Tampa	12842	2758	8232	7.6	85	100	100	101	76	67	40	38	61	65	73	80	68
FL	Tampa, Macdill AFB	12810	2751	8230	6.9	73	95	98	83	59	51	35	40	67	73	62	67	67
FL	Avon Park Range AAF	12804	2738	8120	5.4	50	51	55	64	45	30	18	24	61	73	43	48	45
FL	Orlando, Herndon APT	12841	2833	8120	8.2	86	110	131	120	99	83	69	85	91	107	89	99	97
FL	Orlando, McCoy AFB	12815	2827	8118	5.9	61	76	71	67	46	41	29	24	43	48	46	51	49
FL	Titusville	109	2831	8047	6.7	57	72	72	58	44	42	43	32	47	56	50	56	49
FL	Cocoa Beach, Patrick AFB	12867	2814	8036	8.8	127	149	144	134	115	80	52	62	130	191	143	127	119
FL	Cape Kennedy AFS	12868	2829	8033	7.4	82	107	103	90	71	55	41	37	82	94	75	76	73
FL	Daytona Beach APT	12834	2911	8103	8.9	112	141	146	142	125	94	91	95	113	161	108	116	120
FL	Jacksonville, Cecil Fld NAS	93832	3013	8157	5.2	43	65	56	50	35	31	21	19	39	39	37	39	39
FL	Jacksonville NAS	93837	3014	8141	6.9	61	80	81	70	58	60	40	38	76	77	62	64	63
FL	Mayport NAAS	3853	3023	8125	7.2	82	105	92	90	67	67	40	39	110	90	74	68	76
FL	Tallahassee	93805	3023	8422	5.8	51	59	76	66	41	28	24	28	39	43	51	51	45
FL	Marianna	13851	3050	8511	6.9	92	104	115	86	65	48	43	36	55	61	71	84	72
FL	Panama City, Tyndall AFB	13846	3004	8535	6.7	79	101	120	97	62	47	42	37	65	55	64	75	71
FL	Crestview	13884	3047	8631	5.6	68	77	85	57	31	22	16	16	35	38	60	65	47
FL	Valparaiso, Eglin AFB	13858	3029	8631	6.2	66	74	78	71	56	48	40	37	55	46	55	59	56
FL	Valparaiso, Duke Fld	3844	3039	8632	7.0	104	123	105	115	78	46	33	38	40	48	84	88	72
FL	Valparaiso, Hurlburt Fld	3852	3025	8641	5.5	55	62	55	51	36	31	23	21	34	33	39	45	40
FL	Milton, Whiting Fld NAAS	93841	3042	8701	7.1	107	114	125	93	62	44	36	32	65	57	84	92	73
FL	Pensacola, Saufley Fld NAS	3815	3026	8711	6.8	98	109	110	94	57	42	37	35	79	63	81	99	74
FL	Pensacola, Ellyson Fld	3840	3032	8712	7.8	87	104	116	112	86	62	48	44	65	57	74	81	74
FL	Pensacola, Forest Sherman Fd	3855	3021	8719	8.0	110	119	113	106	79	75	56	57	73	71	86	99	88
GA	Valdosta, Moody AFB	13857	3058	8312	4.8	40	51	54	43	29	28	21	19	33	32	29	35	37
GA	Moultrie	13835	3108	8342	6.6	73	89	84	79	45	32	34	30	47	60	59	75	58
GA	Albany, Turner AFB	13815	3135	8407	5.3	50	68	73	55	33	27	23	19	33	27	36	41	41
GA	Brunswick, Glynco NAS	93836	3115	8128	5.5	39	54	53	52	40	36	28	25	38	39	35	37	40
GA	Ft. Stewart, Wright AAF	3875	3153	8134	3.7	20	30	32	23	22	14	12	10	14	16	15	23	20
GA	Savannah	3822	3208	8112	7.5	88	108	98	87	56	49	46	45	61	62	63	72	69
GA	Savannah, Hunter AFB	13824	3201	8108	5.8	59	76	86	70	44	40	34	31	38	43	48	47	51

Monthly Average Wind Power in the United States and Southern Canada (Continued)

State	Location	Sta. No.	Lat	Long	Ave. Speed knots	J	F	M	A	M	J	J	A	S	O	N	D	Ave.
GA	Macon	13836	3242	8339	8.0	92	112	103	117	69	59	56	44	61	56	68	73	75
GA	Warner Robbins AFB	13860	3238	8336	4.9	54	76	75	59	35	26	22	18	27	31	42	44	41
GA	Ft. Benning	13829	3221	8500	3.9	45	64	71	51	28	21	13	13	21	22	32	36	33
GA	Winder	12	3400	8342	7.6	99	113	93	91	57	50	51	44	43	79	92	92	78
GA	Adairsville	110	3455	8456	6.2	87	95	109	74	56	42	36	33	33	49	100	71	64
GA	Augusta, Bush Fld	3820	3322	8158	5.9	68	83	87	83	43	41	36	32	43	39	45	49	53
GA	Atlanta	13874	3339	8426	8.5	170	169	165	151	84	67	56	46	73	80	109	127	106
GA	Marietta, Dobbins AFB	13864	3355	8432	5.8	89	99	105	96	52	38	34	30	40	50	66	72	66
HI	Honolulu IAP	22521	2120	15055	9.8	118	131	164	163	155	172	189	194	141	128	133	144	153
HI	Barbers Point NAS	22514	2119	15804	8.3	106	99	104	102	93	97	100	102	77	76	95	104	95
HI	Wahiawa, Wheeler AFB	22508	2129	15802	5.9	48	49	61	59	60	70	72	65	43	40	39	49	54
HI	Waialua, Mokoleia Fld	22507	2135	15812	7.7	59	52	97	141	115	136	151	158	113	84	89	108	109
HI	Kaneohe Bay MCAS	22519	2127	15747	10.0	131	144	157	156	140	137	143	143	116	113	135	168	141
HI	Barking Sands AAF	22501	2203	15947	5.6	112	62	42	40	33	24	20	22	21	38	43	69	44
HI	Molokai, Homestead Fld	22502	2109	15706	12.3	110	195	249	291	250	312	361	342	266	268	233	238	266
HI	Kahului NAS	22516	2054	15626	11.1	203	204	240	276	335	366	375	377	283	219	247	200	276
HI	Hilo	21504	1943	15504	7.7	82	86	77	71	65	67	63	67	59	56	52	74	67
ID	Strevell	179	4201	11313	9.7	275	255	189	175	161	148	127	120	127	128	188	209	168
ID	Pocatello	24156	4255	11236	8.6	211	224	230	209	163	159	113	87	102	103	148	176	160
ID	Idaho Falls	671	4331	11204	9.7	226	185	321	295	241	214	132	139	166	184	178	172	200
ID	Burley APT	24133	4232	11346	8.0	185	162	246	199	156	116	73	58	72	89	114	150	133
ID	Twin Falls	15634	4228	11429	8.7	168	181	232	237	155	139	85	74	86	114	131	172	147
ID	King Hill	185	4259	11513	8.8	220	222	357	363	330	216	169	147	212	158	165	185	221
ID	Mountain Home AFB	24106	4303	11552	7.3	94	136	154	172	144	121	92	76	80	105	89	81	110
ID	Boise APT	24131	4334	11613	7.8	103	112	127	119	95	79	64	56	60	76	84	95	91
IL	Chicago Midway	14819	4147	8745	9.0	129	145	151	144	115	70	53	52	74	95	149	134	112
IL	Glenview NAS	14855	4205	8750	8.4	164	164	203	206	137	83	56	52	72	105	143	137	128
IL	Chicago, Ohare	14810	4159	8754	9.7	220	242	268	272	197	140	99	89	140	162	258	213	193
IL	Chicago, Ohare IAP	94846	4159	8754	9.5	189	199	227	229	174	118	83	71	113	129	227	176	162
IL	Waterman	139	4146	8845	9.1	236	269	222	269	134	100	50	60	77	99	210	175	166
IL	Rockford	94822	4212	8906	8.8	112	107	135	164	126	85	61	70	82	92	126	121	107
IL	Moline	14923	4127	9031	8.9	121	151	215	200	155	93	63	54	91	113	185	141	130
IL	Bradford	146	4113	8937	10.2	210	271	284	290	203	129	68	88	96	123	237	183	196
IL	Rantoul, Chanute AFB	14806	4018	8809	8.5	158	164	193	210	143	91	50	47	66	87	145	127	121
IL	Effingham	436	3909	8832	9.3	170	210	251	217	116	95	73	68	89	95	217	144	136
IL	Springfield, Capitol APT	93822	3950	8940	10.6	215	253	308	295	212	131	92	82	119	152	263	242	198
IL	Quincy, Baldwin Fld	93989	3956	9112	9.9	209	229	275	220	137	98	71	61	95	136	211	194	161
IL	Belleville, Scott AFB	13802	3833	8951	7.2	129	140	162	143	85	61	36	33	46	56	109	96	90
IL	Marion, Williamson Co APT	3865	3745	8901	7.6	159	186	230	243	136	88	59	45	88	93	187	28	139
IN	Evansville	93817	3808	8732	8.1	129	139	165	154	98	69	46	38	60	71	118	117	100
IN	Terre Haute, Holman Fld	3868	3927	8717	8.2	160	151	203	182	105	74	43	33	60	79	130	134	115
IN	Indianapolis	93819	3944	8617	7.1	174	198	247	205	147	96	68	59	81	108	178	161	143
IN	Columbus, Bakalar AFB	13803	3916	8554	7.0	97	106	128	117	71	50	36	32	44	58	91	88	74
IN	Milroy	125	3928	8522	9.5	243	270	230	209	116	115	73	67	94	101	189	163	148
IN	Centerville	130	3949	8458	9.0	196	237	209	182	101	87	64	57	79	93	176	136	134

State	Location	Sta. No.	Lat	Long	Ave. Speed knots	J	F	M	A	M	J	J	A	S	O	N	D	Ave.
IN	Marion APT	94852	4029	8541	8.4	211	254	279	255	160	116	64	50	79	95	253	186	170
IN	Peru, Grissom AFB	94833	4039	8609	7.7	123	137	158	165	109	65	40	36	53	69	131	133	100
IN	Lafayette	530	4025	8656	10.3	290	317	290	316	175	142	91	98	112	126	296	222	215
IN	Fort Wayne	14827	4100	8512	9.5	149	167	230	205	154	101	73	66	96	116	214	171	146
IN	Helmer	535	4133	8512	9.6	256	243	263	242	141	100	79	78	133	153	242	215	161
IN	Goshen	132	4132	8548	8.9	229	209	208	221	124	104	75	73	89	103	182	148	146
IN	South Bend	138	4142	8619	9.8	243	243	283	256	155	128	92	92	107	127	229	156	188
IN	McCool	136	4133	8710	10.7	284	297	311	290	183	149	82	96	130	157	311	231	223
IA	Dubuque APT	94908	4224	9042	9.4	208	209	239	317	241	150	112	135	170	198	312	231	210
IA	Burlington	14931	4046	9107	9.5	147	160	257	164	94	85	52	44	72	97	200	144	128
IA	Iowa City APT	14937	4138	9133	8.6	175	195	231	229	118	82	71	61	81	100	205	159	146
IA	Cedar Rapids	14990	4153	9142	9.2	160	171	23	249	157	97	53	49	59	102	138	131	132
IA	Ottumwa	14948	4106	9226	9.1	209	243	257	239	169	140	118	112	156	169	174	168	179
IA	Montezuma	145	4135	9228	11.0	270	330	330	390	256	203	103	117	150	158	271	223	237
IA	Des Moines	14933	4132	9339	9.9	182	193	251	289	180	126	81	81	109	142	219	178	168
IA	Ft. Dodge APT	94933	4233	9411	10.3	253	260	331	334	258	140	77	74	104	181	185	188	199
IA	Atlantic	140	4122	9503	11.3	296	350	363	457	295	256	136	123	155	190	264	270	256
IA	Sioux City	14943	4224	9623	9.7	180	172	247	283	206	143	89	82	114	155	212	170	169
KS	Ft. Leavenworth	13921	3922	9455	6.3	73	84	116	111	73	57	31	33	50	52	78	66	69
KS	Olathe NAS	93909	3850	9453	9.2	143	157	211	187	139	117	69	69	91	102	152	136	135
KS	Topeka	13996	3904	9538	9.8	138	147	237	229	170	159	107	112	136	137	159	163	157
KS	Topeka, Forbes AFB	13920	3857	9540	8.6	117	134	186	185	132	115	69	79	88	95	125	104	120
KS	Ft. Riley	13947	3903	9646	8.0	112	122	224	233	171	130	86	102	139	138	125	106	139
KS	Cassoday	152	3802	9638	13.0	370	436	550	550	350	310	231	257	283	284	371	311	377
KS	Wichita	3928	3739	9725	12.0	243	273	344	337	262	276	168	177	203	221	249	237	253
KS	Wichita, McConnell AFB	3923	3737	9716	10.9	222	234	336	317	252	237	151	136	176	188	200	207	222
KS	Hutchinson	93905	3756	9754	10.7	287	335	372	375	330	351	215	195	309	280	308	269	305
KS	Salina, Schilling AFB	13922	3848	9738	9.1	134	168	230	221	176	150	100	112	148	135	147	111	155
KS	Hill City APT	93990	3923	9950	9.7	122	199	337	262	210	226	152	125	153	140	153	131	184
KS	Dodge City APT	13985	3746	9958	13.5	281	360	441	458	360	368	259	245	296	296	334	318	336
KS	Garden City APT	23064	3756	10043	12.4	277	326	451	450	415	456	277	272	309	259	216	204	295
KY	Corbin	3814	3658	8408	4.4	71	54	65	58	26	15	16	12	16	18	43	44	36
KY	Lexington	93820	3802	8436	8.9	161	158	156	169	103	75	61	47	72	73	148	146	113
KY	Warsaw	23	3846	8454	6.8	123	132	137	123	65	57	49	39	40	54	106	101	85
KY	Louisville, Standiford Fld	93821	3811	8544	6.5	75	84	104	96	53	32	26	27	29	36	58	66	56
KY	Ft. Knox	13807	3754	8558	6.6	108	121	126	111	64	46	30	25	39	47	98	96	76
KY	Bowling Green, City Co APT	93808	3658	8626	6.6	131	115	136	113	65	39	38	32	46	57	89	93	79
KY	Ft. Campbell	13806	3640	8730	5.8	78	88	107	89	51	32	27	25	29	39	60	69	56
KY	Paducah	3816	3704	8846	6.7	109	106	122	106	59	44	35	33	40	48	90	93	74
LA	New Orleans	12916	2959	9015	8.0	129	137	144	114	76	52	44	43	81	91	128	109	97
LA	New Orleans, Callender NAS	12958	2949	9001	4.6	47	55	50	35	26	14	10	10	26	24	31	40	30
LA	Baton Rouge	13970	3032	9109	7.4	105	106	102	95	72	52	40	36	50	53	79	92	73
LA	Lake Charles, Chenault AFB	13941	3013	9310	8.3	184	156	204	176	125	91	58	57	67	67	133	140	120
LA	Polk AAF	3931	3103	9311	5.7	51	69	78	68	47	37	23	15	21	29	55	51	47
LA	Alexandria, England AFB	13934	3119	9233	4.6	45	57	64	52	37	20	15	13	17	21	39	41	34

Monthly Average Wind Power in the United States and Southern Canada (Continued)

State	Location	Sta. No.	Lat	Long	Ave. Speed knots	J	F	M	A	M	J	J	A	S	O	N	D	Ave.
LA	Monroe, Selman Fld	13942	3231	9203	7.0	88	104	108	90	61	46	36	36	46	51	73	79	67
LA	Shreveport	13957	3228	9349	8.4	128	138	145	131	92	71	57	55	58	69	105	111	98
LA	Shreveport, Barksdale AFB	13944	3230	9340	6.0	69	74	83	72	48	36	27	27	35	34	53	59	50
ME	Portland	14764	4339	7019	8.4	127	145	158	140	103	80	197	61	83	101	112	120	107
ME	Brunswick, NAS	14611	4353	6956	6.8	109	116	106	107	87	64	55	48	58	69	80	97	82
ME	Bangor, Dow AFB	14601	4448	6841	7.1	132	138	136	113	83	70	54	59	63	82	100	110	93
ME	Presque Isle AFB	14604	4641	6803	7.8	151	167	151	161	123	88	77	69	97	115	110	134	120
ME	Limestone, Loring AFB	14623	4657	6753	6.9	97	107	110	88	69	55	48	45	60	68	73	78	74
MD	Patuxent River NAS	13721	3817	7625	8.1	159	177	186	148	97	76	59	59	83	102	138	139	119
MD	Baltimore, Martin Fld	93744	3920	7625	6.9	107	111	119	95	53	44	37	39	34	46	57	64	61
MD	Baltimore, Friendship APT	93721	3911	7640	9.6	206	253	265	209	152	117	96	79	110	117	179	188	164
MD	Ft. Mead, Tipton AAF	93733	3905	7646	4.4	57	58	69	65	37	19	14	14	14	23	42	41	38
MD	Aberdeen, Phillips AAF	13701	3928	7610	7.9	126	170	173	157	95	66	52	55	69	95	121	118	109
MD	Camp Detrick, Frederick	13749	3926	7727	5.4	101	122	144	110	51	33	25	22	30	47	96	75	72
MD	Ft. Ritchie	93745	3944	7724	4.6	38	34	33	37	21	16	14	23	18	27	27	52	28
MA	Chicopee Falls, Westover AAF	14703	4212	7232	7.1	122	143	131	133	96	70	52	48	60	81	104	114	96
MA	Ft. Devons AAF	4779	4234	7136	5.4	45	40	66	84	44	31	29	32	33	39	48	53	45
MA	Bedford, Hanscom Fld	14702	4228	7117	6.1	109	120	117	94	70	48	39	36	44	65	80	92	76
MA	Boston, Logan IAP	14739	4222	7102	11.8	314	321	314	268	195	150	128	108	131	131	230	277	227
MA	Boston	14739	4222	7102	11.7	314	321	314	268	195	150	128	108	131	131	230	277	227
MA	South Weymouth NAS	14790	4209	7056	7.6	125	125	146	136	84	58	43	56	53	71	92	102	90
MA	Falmouth, Otis AFB	14704	4139	7031	9.2	188	199	198	193	147	110	87	90	112	139	148	185	149
MA	Nantucket	14756	4116	7003	11.6	304	346	298	277	190	140	104	113	169	214	261	298	223
MA	Nantucket Shoals	14658	4101	6930	16.7	1024	1025	977	838	632	551	592	544	482	769	856	927	757
MA	Georges Shoals	14657	4141	6747	17.1	1168	1175	1058	891	619	575	519	378	473	739	891	1156	783
MI	Mt. Clemens, Selfridge AFB	14804	4236	8249	8.2	157	151	160	145	96	71	56	53	71	84	156	144	115
MI	Ypsilanti, Willow Run	14853	4214	8332	9.5	169	169	244	194	139	101	85	77	104	113	188	173	147
MI	Jackson	133	4216	8428	8.8	196	149	182	215	106	92	57	69	77	99	175	147	127
MI	Battle Creek, Kellogg APT	14815	4218	8514	8.9	161	189	205	179	124	99	76	63	106	99	152	180	137
MI	Grand Rapids	94860	4253	8531	8.7	120	134	180	158	112	77	62	54	83	87	162	135	113
MI	Lansing	14836	4247	8436	10.8	273	298	356	287	178	112	69	74	123	146	251	269	203
MI	Flint, Bishop APT	14826	4258	8344	9.6	233	206	246	195	140	109	85	71	129	140	210	223	167
MI	Saginaw, Tri City APT	14845	4326	8352	9.7	218	196	223	196	152	111	92	74	121	128	199	189	158
MI	Muskegon Co APT	14840	4310	8614	9.4	156	164	140	171	121	96	68	70	80	155	177	166	129
MI	Gladwin	14828	4359	8429	5.9	67	71	92	76	63	40	31	24	34	39	61	53	53
MI	Cadillac APT	14817	4415	8528	9.4	210	204	239	193	172	151	104	89	139	161	208	203	171
MI	Traverse City	14850	4444	8535	9.5	229	206	249	207	147	132	100	91	160	187	250	225	178
MI	Oscoda, Wurtsmith AFB	14808	4427	8322	7.6	117	123	121	116	91	76	55	58	71	94	109	108	94
MI	Alpena, Collins Fld	94849	4504	8334	7.3	76	76	92	108	88	59	49	46	52	62	67	58	70
MI	Pellston, Emmett Co APT	14841	4534	8448	8.9	183	159	192	165	154	115	105	81	115	144	175	185	147
MI	Sault Ste Marie	14847	4628	8422	8.3	114	105	119	125	113	77	62	57	79	93	115	108	98
MI	Kinross, Kincheloe AFB	94824	4615	8428	7.6	88	105	106	120	106	69	53	56	68	81	109	93	89
MI	Escanaba APT	94853	4544	8705	7.8	126	164	148	186	191	150	116	93	135	163	232	143	154
MI	Gwinn, Sawyer AFB	94836	4621	8723	7.5	94	116	109	116	100	72	52	57	65	92	102	105	90
MI	Marquette	14838	4634	8724	7.6	78	84	118	125	117	96	73	70	89	85	88	66	91

State	Location	Sta. No.	Lat	Long	Ave. Speed knots	J	F	M	A	M	J	J	A	S	O	N	D	Ave.
MI	Houghton Co APT	14858	4710	8830	8.5	116	126	136	139	106	90	78	66	97	108	116	112	108
MI	Ironwood, Gogebic Co APT	94926	4632	9008	8.5	164	198	167	290	280	174	130	139	200	213	270	203	202
MN	Minneapolis, St. Paul IAP	14922	4453	9313	9.4	127	142	152	211	186	133	88	86	112	130	167	123	138
MN	St. Cloud, Whitney APT	14926	4535	9411	6.9	70	70	102	129	99	71	44	38	55	66	84	58	74
MN	Alexandria	751	4553	9524	10.7	221	215	262	290	249	202	128	161	182	263	263	196	229
MN	Brainerd	94938	4624	9408	6.9	90	92	111	167	134	98	62	58	107	87	123	89	102
MN	Duluth IAP	14913	4650	9211	10.7	219	229	249	299	233	147	122	111	154	196	254	206	198
MN	Bemidji APT	14958	4730	9456	7.2	104	120	111	244	201	155	117	120	133	141	162	122	144
MN	International Falls APT	14918	4834	9323	8.4	95	103	110	175	155	103	82	88	119	122	163	117	119
MN	Roseau	14955	4851	9545	6.5	33	31	50	58	51	35	16	20	27	34	49	41	37
MN	Thief River Falls	8243	4803	9611	8.7	215	207	194	305	279	199	135	157	189	221	309	219	218
MS	Biloxi, Keesler AFB	13820	3024	8855	6.8	82	79	83	81	64	49	38	35	58	55	66	68	63
MS	Jackson	13927	3220	9014	6.2	85	92	88	78	46	31	26	25	31	39	63	77	56
MS	Greenville APT	13939	3329	9059	6.6	89	100	104	90	66	48	32	34	49	50	65	77	67
MS	Meridian NAAS	3866	3323	8833	3.5	29	40	37	25	12	8	9	5	8	10	17	21	18
MS	Columbus AFB	13825	3338	8827	4.7	52	60	61	48	25	17	15	13	23	21	31	40	34
MO	Malden	13848	3636	8959	8.3	151	117	162	152	109	75	57	53	62	74	124	118	106
MO	St. Louis, Lambert Fld	13994	3845	9023	7.9	95	116	143	138	94	60	41	36	55	60	92	96	83
MO	New Florence	147	3853	9126	10.1	198	231	238	231	138	105	85	85	111	112	199	165	158
MO	Kirksville	540	4006	9232	10.5	250	271	297	310	191	137	111	103	118	158	231	191	184
MO	Vichy, Rolla APT	13997	3808	9146	8.6	153	158	211	170	92	72	54	45	68	76	142	156	117
MO	Ft. Leonard Wood, Forney AF	3938	3743	9208	6.0	67	65	81	88	50	37	22	21	26	50	62	67	52
MO	Springfield	440	3714	9315	9.7	183	243	230	263	123	96	70	77	97	10	183	170	150
MO	Butler	93995	3818	9420	9.3	212	208	266	226	123	124	74	65	81	131	156	160	152
MO	Knobnoster, Whiteman AFB	13930	3844	9334	7.4	96	109	146	151	94	65	40	45	61	70	94	81	87
MO	Marshall	144	3906	9312	9.5	203	223	263	250	115	109	82	90	89	95	156	136	143
MO	Grandview, Rchds-Gebaur AFB	3929	3851	9435	8.1	105	101	156	173	113	76	55	59	74	91	110	110	100
MO	Kansas City APT	13988	3907	9436	9.4	115	126	165	182	145	129	105	99	113	112	143	122	132
MO	Knoxville	142	3925	9400	10.1	210	211	284	278	144	124	84	91	98	125	171	151	158
MO	Tarkio	14945	4027	9522	8.2	121	137	258	225	192	150	87	71	85	113	135	94	135
MT	Glendive	24087	4708	10448	7.7	131	141	145	222	217	146	125	137	139	144	111	122	149
MT	Miles City APT	24037	4626	10552	8.6	123	137	120	161	134	102	87	98	104	109	93	116	115
MT	Wolf Point	94017	4806	10535	8.1	117	112	143	326	248	139	126	171	245	223	183	124	179
MT	Glasgow AFB	94010	4824	10631	8.6	133	131	125	178	198	130	102	101	138	126	119	125	133
MT	Billings, Logan Fld	24033	4548	10832	10.0	230	210	185	202	165	137	110	99	128	152	218	237	173
MT	Livingston	678	4540	11032	13.5	778	819	574	415	327	239	233	253	321	500	713	1058	500
MT	Lewiston APT	24036	4703	10927	8.6	198	185	141	163	135	108	82	95	114	125	189	153	140
MT	Havre	777	4834	10940	8.7	148	106	155	141	123	115	75	74	86	120	132	127	114
MT	Great Falls IAP	24143	4729	11122	11.6	444	439	300	281	194	193	136	143	197	300	456	509	304
MT	Great Falls, Malmstrom AFB	24112	4731	11110	8.9	253	240	181	176	120	112	80	80	115	178	215	263	169
MT	Helena APT	24144	4636	11200	7.3	145	95	142	134	63	113	85	44	111	34	61	65	90
MT	Whitehall	24161	4552	11158	11.4	710	543	352	274	221	245	193	167	174	260	410	602	344
MT	Butte, Silver Bow Co APT	24135	4557	11230	6.9	98	101	116	158	141	120	86	85	93	93	86	76	104
MT	Missoula	24153	4655	11405	5.0	49	36	64	75	72	70	64	41	57	23	19	26	50
NB	Omaha	14942	4118	9554	10.0	191	186	264	280	186	148	104	104	122	158	217	191	177

Monthly Average Wind Power in the United States and Southern Canada   (Continued)

State	Location	Sta. No.	Lat	Long	Ave. Speed knots	J	F	M	A	M	J	J	A	S	O	N	D	Ave.
NB	Omaha, Offutt AFB	14949	4107	9555	7.6	118	123	189	198	138	98	67	58	68	93	112	112	115
NB	Grand Island APT	14935	4058	9819	11.1	177	195	270	312	251	217	161	158	179	180	232	200	211
NB	Overton	154	4044	9927	10.5	209	195	323	389	276	243	155	149	162	202	299	182	222
NB	North Platte	562	4108	11042	10.5	193	233	374	435	321	208	153	153	206	246	234	166	254
NB	Lincoln AFB	14904	4051	9646	9.4	163	173	258	251	193	143	97	102	102	119	173	146	162
NB	Columbus	73084	4126	9720	10.0	184	192	316	301	246	172	113	111	120	184	150	143	186
NB	Norfolk, Stefan APT	14941	4159	9726	9.7	236	235	308	387	275	215	142	173	204	281	361	255	256
NB	Big Springs	161	4105	10207	11.7	270	284	430	450	349	270	210	210	217	264	297	251	290
NB	Sidney	563	4108	10302	10.5	275	267	369	395	294	227	193	160	680	227	241	188	248
NB	Scottsbluff APT	24028	4152	10336	9.8	147	225	271	254	189	180	119	124	122	166	254	199	165
NB	Alliance	24044	4203	10248	10.6	203	209	274	358	289	233	189	204	233	228	238	210	238
NB	Valentine, Miller Fld	24032	4252	10033	10.0	181	237	267	323	286	242	199	226	230	271	338	242	253
NV	Boulder City	382	3558	11450	7.6	109	162	185	230	247	293	186	197	137	95	155	81	173
NV	Las Vegas	23169	3605	11510	8.7	105	142	186	229	225	209	166	141	111	122	67	92	150
NV	Las Vegas, Nellis AFB	23112	3615	11502	5.7	73	89	137	138	125	123	77	75	61	63	66	57	88
NV	Indian Springs AFB	23141	3635	11541	5.3	38	75	141	196	154	109	64	46	54	28	63	54	80
NV	Tonopah APT	23153	3804	11708	8.7	99	150	196	196	174	142	98	94	104	113	109	101	133
NV	Fallon NAAS	93102	3925	11843	4.7	48	50	74	72	60	49	29	23	25	30	27	36	44
NV	Reno	23185	3930	11947	5.2	77	108	123	99	93	81	56	54	52	52	45	44	74
NV	Reno, Stead AFB	23118	3940	11952	5.9	79	105	125	132	110	91	72	72	56	64	51	69	85
NV	Humboldt	580	4005	11809	6.7	61	76	140	102	103	118	98	83	66	57	47	48	79
NV	Lovelock	24172	4004	11833	6.4	109	91	121	99	98	113	80	72	56	67	43	47	83
NV	Winnemucca APT	24128	4054	11748	7.2	79	91	117	115	105	97	89	81	73	80	55	60	86
NV	Buffalo Valley	24181	4020	11721	6.4	66	97	94	94	102	97	77	62	57	59	53	52	76
NV	Battle Mountain	24119	4037	11652	7.3	148	80	147	113	132	113	85	70	61	102	64	66	98
NV	Beowawe	181	4036	11631	6.2	57	98	112	106	99	86	79	68	63	62	45	51	76
NV	Elko	582	4050	11548	6.2	68	76	100	92	99	98	91	76	75	73	52	60	76
NV	Ventosa	70	4052	11448	6.8	139	160	178	179	166	112	104	97	96	88	93	99	109
NH	Portsmouth, Pease AFB	4743	4305	7049	6.6	90	112	99	82	73	50	39	37	42	54	63	90	68
NH	Manchester, Grenier Fld	14710	4256	7126	6.7	105	150	134	137	83	68	49	37	51	72	95	127	86
NH	Keene	94721	4254	7216	4.8	63	86	69	74	67	48	29	31	34	42	43	50	53
NJ	Atlantic City	93730	3927	7435	9.1	185	207	207	166	109	81	62	61	81	107	144	164	129
NJ	Camden	103	3955	7504	8.0	132	131	167	160	86	73	64	55	62	82	118	111	104
NJ	Wrightstown, McGuire AFB	14706	4000	7436	6.6	100	114	114	101	48	42	30	28	40	52	73	89	69
NJ	Lakehurst NAS	14780	4002	7420	7.4	133	158	166	131	94	62	49	41	48	60	99	109	93
NJ	Belmar	4739	4011	7404	6.1	80	82	83	57	37	31	23	23	30	50	56	55	50
NJ	Trenton	501	4017	7450	8.1	125	146	146	194	85	71	47	55	70	92	140	120	105
NJ	Newark	14734	4042	7410	8.7	145	145	157	126	107	80	73	68	71	96	100	110	109
NM	Clayton	23051	3627	10309	13.0	447	397	519	483	427	360	230	207	255	279	350	395	354
NM	Tucumcari	364	3511	10336	10.6	273	321	359	365	293	227	167	154	166	207	206	206	260
NM	Anton Chico	160	3508	10505	8.9	204	257	306	227	130	132	78	67	74	101	145	140	155
NM	Clovis, Cannon AFB	23008	3423	10319	9.9	171	215	320	279	228	204	126	93	111	122	160	180	186
NM	Hobbs, Lea Co APT	93034	3241	10312	10.4	195	234	353	276	250	215	138	109	118	109	166	188	190
NM	Roswell APT	23043	3324	10432	8.5	148	191	273	260	216	172	101	82	84	102	126	172	163
NM	Roswell, Walker AFB	23009	3318	10432	7.3	79	102	145	145	129	135	93	71	65	72	82	86	98

State	Location	Sta. No.	Lat	Long	Ave. Speed knots	J	F	M	A	M	J	J	A	S	O	N	D	Ave.
NM	Rodeo	272	3156	10859	9.4	195	216	250	325	259	191	166	131	129	155	210	173	203
NM	Las Cruces, White Sands	23039	3222	10629	6.1	100	106	169	149	123	88	50	43	41	42	82	99	89
NM	Alamogordo, Holloman AFB	23002	3251	10605	5.6	45	59	92	101	85	72	54	43	38	35	41	40	57
NM	Albuquerque, Kirtland AFB	23050	3503	10637	7.6	83	115	154	190	160	134	101	72	88	97	80	74	112
NM	Otto	166	3505	10600	9.6	248	311	491	372	271	264	116	102	94	176	234	228	243
NM	Santa Fe APT	23049	3537	10605	10.3	218	200	308	308	248	217	138	113	135	154	184	194	201
NM	Farmington APT	23090	3645	10814	7.1	53	74	136	151	106	101	78	55	50	71	81	42	83
NM	Gallup	23081	3531	10847	6.2	92	133	237	293	248	217	92	82	75	114	84	54	143
NM	Zuni	93044	3506	10848	8.4	127	109	234	220	183	138	58	54	80	104	97	126	127
NM	El Morro	373	3501	10826	7.4	76	113	229	210	185	136	95	66	66	92	91	83	107
NM	Acomita	170	3503	10743	9.6	150	169	283	223	156	143	96	82	75	109	143	136	169
NY	Westhampton, Suffolk Co AFB	14719	4051	7238	8.1	146	145	154	133	100	82	69	67	87	110	120	118	110
NY	Hempstead, Mitchell AFB	14708	4044	7336	9.2	194	221	211	189	134	115	99	88	100	129	185	194	155
NY	New York, Kennedy IAP	94789	4039	7347	10.3	242	259	260	204	151	139	122	106	120	140	173	180	168
NY	New York, La Guardia	14732	4046	7354	10.9	300	282	283	211	160	123	105	110	135	174	217	278	197
NY	New York, Central Park	94728	4047	7358	8.1	114	108	117	99	57	43	38	37	61	63	83	95	76
NY	New York WBO	94706	4043	7400	11.6	436	428	384	259	211	173	146	107	143	209	329	336	261
NY	Bear Mountain	100	4114	7400	12.5	476	463	550	444	311	183	171	163	271	289	396	523	350
NY	Newburgh, Stewart AFB	14714	4130	7406	7.8	164	206	193	176	108	76	61	52	64	101	136	163	124
NY	New Hackensack	106	4138	7353	6.0	83	91	93	87	50	42	34	32	40	63	93	84	63
NY	Poughkeepsie, Duchess Co APT	14757	4138	7353	6.1	68	90	106	90	52	43	33	28	36	46	66	74	60
NY	Columbiaville	115	4220	7345	8.7	185	220	226	173	131	104	69	69	97	138	164	172	138
NY	Albany Co APT	14735	4245	7348	7.9	148	163	173	138	95	80	68	63	81	96	103	111	108
NY	Schenectady	4782	4251	7357	7.4	156	116	160	155	123	81	82	67	84	68	112	114	112
NY	Plattsburg AFB	4742	4439	7327	6.0	63	78	76	82	70	52	42	36	41	54	66	60	60
NY	Massena, Richards APT	94725	4456	7451	9.5	176	192	217	193	150	129	108	101	111	154	170	193	158
NY	Watertown APT	94790	4400	7601	10.0	409	312	373	298	153	134	119	99	171	197	278	350	236
NY	Rome, Griffiss AFB	14717	4314	7525	5.7	91	108	109	94	66	44	30	26	36	50	71	82	65
NY	Utica, Oneida Co APT	94794	4309	7523	8.2	117	130	113	100	74	81	43	49	59	67	106	111	87
NY	Syracuse, Hancock APT	14771	4307	7607	8.4	158	174	166	162	108	80	66	61	76	91	129	138	115
NY	Binghamton, Bloome Co APT	4725	4213	7559	9.0	157	183	194	191	138	77	73	70	77	104	155	160	132
NY	Elmira, Chemung Co APT	14748	4210	7654	5.6	73	78	91	80	50	45	29	25	34	57	79	69	59
NY	Rochester	14768	4307	7740	9.8	205	229	240	201	138	123	98	82	102	123	197	194	163
NY	Buffalo	528	4256	7843	11.5	468	430	417	411	422	281	278	383	377	334	354	306	382
NY	Buffalo	14733	4256	7849	10.9	254	258	322	239	160	151	132	118	145	160	227	251	205
NY	Niagara Falls	4724	4306	7857	8.3	193	175	147	130	106	82	72	67	82	105	134	176	126
NY	Dunkirk	127	4230	7916	11.2	488	348	368	302	173	154	121	127	174	269	396	361	281
NC	Wilmington	13748	3416	7755	8.1	118	151	163	169	102	87	77	80	98	97	101	97	108
NC	Jacksonville, New Rvr. MCAF	93727	3443	7726	6.0	59	72	84	79	53	44	32	31	43	40	49	46	51
NC	Cherry Point NAS	13754	3454	7653	7.0	92	105	124	125	84	68	57	57	84	66	67	74	83
NC	Cape Hatteras	93729	3516	7533	10.6	195	229	209	202	144	138	117	135	180	160	166	168	169
NC	Goldsboro, Symr-Jhnsn AFB	13713	3520	7758	5.4	55	71	80	72	45	32	30	23	30	29	42	46	45
NC	Ft. Bragg, Simmons AAF	93737	3508	7856	5.8	63	82	76	70	46	32	27	24	27	33	53	48	46
NC	Fayetteville, Pope AFB	13714	3512	7901	4.3	43	54	60	55	34	25	24	21	21	23	29	30	33
NC	Charlotte, Douglas APT	13881	3513	8056	7.4	101	101	120	118	69	57	50	53	70	76	78	82	82

Monthly Average Wind Power in the United States and Southern Canada (Continued)

State	Location	Sta. No.	Lat	Long	Ave. Speed knots	J	F	M	A	M	J	J	A	S	O	N	D	Ave.
NC	Asheville	13872	3536	8232	5.5	77	77	110	90	41	24	17	16	18	33	76	74	54
NC	Hickory APT	3810	3545	8123	7.2	69	69	89	79	57	50	50	49	49	54	63	61	62
NC	Winston Salem	93807	3608	8014	8.1	141	166	149	169	88	68	66	54	97	106	98	117	111
NC	Greensboro	13723	3605	7957	6.7	67	90	94	94	47	37	34	29	35	43	69	57	58
NC	Raleigh	13722	3552	7847	6.7	89	81	106	113	53	51	49	37	45	41	64	61	64
NC	Rocky Mount APT	13746	3558	7748	4.2	72	74	97	86	49	62	51	43	45	50	57	60	52
NC	Elizabeth City	13786	3616	7611	7.4	81	89	95	98	76	65	50	58	67	71	63	63	74
ND	Fargo, Hector APT	14914	4654	9648	11.7	280	264	293	389	286	215	144	160	225	280	337	270	263
ND	Grand Forks AFB	94925	4758	9724	8.9	167	182	183	197	166	103	71	88	123	147	147	172	146
ND	Pembina	758	4857	9715	11.7	308	381	321	341	335	261	187	241	261	329	403	409	308
ND	Bismarck APT	24011	4646	10045	9.5	147	140	186	250	217	174	118	119	157	167	186	143	170
ND	Minot AFB	94011	4825	10121	9.1	191	192	166	199	185	117	95	98	127	164	163	181	157
ND	Williston, Sloulin Fld	94014	4811	10338	8.2	80	86	109	143	141	104	76	83	101	98	88	78	98
ND	Dickinson	24012	4647	10248	13.0	402	365	462	486	401	402	246	208	300	332	426	334	362
OH	Youngstown APT	14852	4116	8040	9.2	187	177	218	178	115	84	66	57	81	95	180	188	133
OH	Warren	21	4117	8048	9.3	196	183	197	197	116	95	67	59	82	122	164	149	136
OH	Akron	14895	4055	8126	9.1	163	184	192	151	101	75	55	55	70	86	156	147	118
OH	Perry	128	4141	8107	10.7	296	296	290	277	136	115	82	90	136	108	311	270	223
OH	Cleveland	14820	4124	8151	10.1	189	237	244	211	147	111	80	72	104	122	230	202	152
OH	Vickery	20	4125	8255	10.7	284	310	303	284	150	136	89	88	130	157	284	217	217
OH	Toledo	94830	4136	8348	7.7	109	114	138	108	76	51	39	37	49	60	89	93	80
OH	Archbold	120	4134	8419	8.8	182	176	182	189	99	86	57	62	84	98	182	135	127
OH	Columbus	14821	4000	8253	7.2	109	118	136	116	74	52	37	35	44	55	100	91	82
OH	Columbus, Lockbourne AFB	13812	3949	8256	6.8	109	127	135	120	70	53	36	33	43	58	93	94	80
OH	Hayesville	124	4047	8218	10.0	257	257	236	204	123	124	76	76	104	151	258	204	163
OH	Cambridge	122	4004	8135	6.2	110	103	110	94	54	50	40	32	40	59	30	73	71
OH	Zanesville, Cambridge	93824	3957	8154	7.7	160	142	188	158	87	67	46	32	56	63	136	130	105
OH	Wilmington, Clinton Co AFB	13841	3926	8348	7.8	133	148	166	157	94	60	44	37	48	63	117	119	93
OH	Cincinnati	93814	3904	8440	8.4	133	135	150	144	90	63	51	42	61	77	126	112	99
OH	Dayton	93815	3954	8413	9.0	179	192	207	173	108	73	59	46	69	84	170	160	125
OH	Dayton, Wright AFB	13813	3947	8406	7.6	161	188	226	182	122	83	58	50	71	86	162	157	128
OH	Dayton, Patterson Fld	13840	3949	8403	7.4	171	186	202	176	108	73	47	43	61	79	160	145	120
OK	Muskogee	13916	3540	9522	8.5	149	189	230	210	132	94	53	71	80	108	114	99	130
OK	Tulsa IAP	13968	3612	9554	9.5	157	178	196	185	155	116	94	85	110	118	145	149	141
OK	Oklahoma City	13967	3524	9736	12.2	306	333	376	386	274	250	165	153	182	216	248	265	263
OK	Oklahoma City, Tinker AFB	13919	3525	9723	11.4	263	277	370	412	319	318	176	153	197	230	243	248	264
OK	Ardmore AFB, Autrey Fld	13903	3418	9701	8.7	140	165	204	203	123	113	71	72	86	98	132	114	127
OK	Ft. Sill	13945	3439	9824	9.2	174	215	272	247	193	182	104	91	125	140	164	165	173
OK	Altus AFB	13902	3439	9916	8.0	99	134	198	180	140	126	71	64	79	92	91	92	113
OK	Clinton-Sherman AFB	3932	3520	9912	9.8	184	202	283	262	224	158	84	77	108	113	139	161	166
OK	Enid, Vance AFB	13909	3620	9754	9.0	162	173	229	188	138	138	88	81	100	107	136	142	139
OK	Waynoka	358	3638	9850	12.4	295	416	562	556	389	310	290	250	275	309	308	261	356
OK	Gage	13975	3618	9946	10.5	203	207	281	323	257	321	168	132	173	161	167	188	221
OR	Ontario	983	4401	11701	6.2	52	70	96	143	139	107	115	122	75	69	49	41	88
OR	Baker	685	4450	11749	7.0	44	52	47	54	47	40	40	40	39	45	38	43	46

State	Location	Sta. No.	Lat	Long	Ave. Speed knots	J	F	M	A	M	J	J	A	S	O	N	D	Ave.
OR	La Grande	24148	4517	11801	8.1	328	261	173	131	93	66	55	54	60	82	201	312	152
OR	Pendleton Fld	24155	4541	11851	8.7	96	164	196	188	174	180	134	124	134	98	127	134	145
OR	Burns	24134	4335	11903	5.9	46	46	68	69	57	60	49	44	46	50	48	40	51
OR	Klamath Falls, Kingsley Fld	94236	4209	12144	4.8	84	77	99	86	62	44	33	30	36	53	66	76	60
OR	Redmond, Roberts Fld	24230	4416	12109	5.6	56	76	63	61	46	36	29	30	38	36	58	45	47
OR	Cascade Locks	192	4539	12150	13.1	651	718	330	331	365	351	387	353	344	451	645	750	465
OR	Crown Point	194	4533	12214	9.6	746	765	209	148	107	50	36	50	113	304	712	650	308
OR	Portland IAP	24229	4536	12236	6.8	139	104	91	61	44	39	45	38	38	51	91	131	75
OR	Eugene, Mahlon Sweet Fld	24221	4407	12313	7.6	83	86	110	94	79	74	89	74	79	58	73	77	81
OR	North Bend	691	4325	12413	8.4	76	128	108	107	88	185	192	149	88	52	80	94	113
OR	Roseburg	690	4314	12321	4.2	22	23	32	26	27	28	29	26	22	16	18	19	23
OR	Astoria, Clatsop Co APT	94224	4609	12353	7.2	125	109	95	82	69	66	71	61	54	70	105	111	84
OR	Salem, McNary Fld	24232	4455	12301	7.1	160	122	100	72	56	47	52	44	47	59	104	137	85
OR	Newport	695	4438	12404	8.5	110	109	107	95	127	145	151	111	69	72	95	129	113
OR	Wolf Creek	87	4241	12323	2.5	10	11	15	16	19	19	22	18	11	9	8	8	14
OR	Sexton Summit	90	4236	12322	11.6	323	283	243	196	236	243	255	269	248	223	316	310	276
OR	Brookings	598	4203	12418	6.4	91	131	96	68	58	55	35	26	37	43	72	92	63
OR	Medford	597	4221	12251	4.9	33	44	53	50	57	51	50	49	39	28	24	40	46
OR	Siskiyou Summit	91	4205	12234	8.8	96	115	102	82	129	150	170	123	102	68	89	82	109
PA	Philadelphia	13739	3953	7515	8.5	131	139	170	133	93	78	62	52	61	84	95	109	103
PA	Willow Grove NAS	14793	4012	7508	6.8	111	136	148	114	74	46	35	30	43	54	88	90	81
PA	Allentown	14737	4039	7526	7.3	157	141	227	124	77	73	38	35	55	61	108	151	104
PA	Scranton	14777	4120	7544	7.7	87	107	94	95	80	62	47	37	50	64	85	84	74
PA	Middletown, Olmstead AFB	14711	4012	7646	5.5	97	124	113	96	53	37	31	28	28	41	75	82	66
PA	Harrisburg	14751	4013	7651	6.4	96	120	125	88	53	41	27	24	31	42	69	68	66
PA	Parkplace	10	4051	7606	12.9	464	477	451	411	251	198	132	137	211	318	411	424	345
PA	Sunbury, Selinsgrove	14770	4053	7646	5.4	72	85	115	73	38	29	19	18	22	32	56	58	50
PA	Woodward	13	4055	7719	13.4	630	564	596	550	330	257	157	163	257	403	551	590	417
PA	Bellefonte	113	4053	7743	6.8	133	139	139	169	83	68	44	46	60	96	128	126	103
PA	Buckstown	114	4004	7850	9.4	261	308	321	241	131	83	73	62	83	179	235	247	192
PA	McConnellsburg	119	3950	7801	7.2	168	148	183	150	92	72	49	42	63	97	174	120	106
PA	Altoona, Blair Co APT	14736	4018	7819	7.9	155	180	241	164	95	79	53	46	56	82	124	143	118
PA	Kylertown	512	4100	7811	9.8	268	241	302	262	147	98	85	77	98	146	202	247	200
PA	Dubois	4787	4111	7854	7.5	118	89	127	118	79	34	36	35	41	49	77	109	78
PA	Bradford	4751	4148	7838	6.1	81	69	76	70	49	30	21	20	25	34	54	68	49
PA	Erie IAP	14860	4205	8011	9.1	234	191	208	147	92	77	67	62	94	111	176	216	139
PA	Mercer	525	4118	8012	9.0	189	169	190	176	109	80	65	58	80	107	170	163	128
PA	Brookville	121	4109	7906	7.4	146	119	119	132	75	60	42	45	46	75	112	104	89
PA	Pittsburg APT	94823	4630	8013	8.5	166	170	187	162	105	75	59	50	67	82	146	150	120
PA	Greensburg	4718	4016	7933	9.1	239	226	207	160	94	75	72	61	80	117	186	174	141
RI	Quonset Point NAS	14788	4135	7125	8.4	164	164	163	156	121	84	61	70	83	115	135	141	123
RI	Providence, Green APT	14765	4144	7126	9.6	173	189	192	180	140	117	98	88	100	117	150	163	144
SC	Beaufort MCAAS	93831	3229	8044	5.7	44	70	62	61	43	35	28	23	35	35	42	43	43
SC	Charleston	13880	3254	8002	7.5	93	124	130	117	69	64	54	52	63	59	71	81	81
SC	Myrtle Beach AFB	13717	3341	7856	6.1	49	65	70	78	54	51	49	44	44	40	39	40	52

Monthly Average Wind Power in the United States and Southern Canada (Continued)

State	Location	Sta. No.	Lat	Long	Ave. Speed knots	J	F	M	A	M	J	J	A	S	O	N	D	Ave.
SC	Florence	300	3411	7943	7.6	95	99	114	93	74	67	52	43	50	74	73	78	72
SC	Sumter, Shaw AFB	13849	3358	8029	5.4	49	57	64	62	40	31	26	24	34	36	38	41	41
SC	Eastover, McIntire ANG	3858	3355	8048	4.9	37	68	54	52	33	24	26	14	32	27	31	28	34
SC	Columbia	13883	3357	8107	6.2	65	74	90	96	52	42	42	34	41	38	44	50	55
SC	Anderson	112	3430	8243	7.7	99	107	99	113	80	60	51	51	52	73	85	98	79
SC	Greenville, Donaldson AFB	13822	3446	8223	6.3	67	70	80	79	44	38	32	27	36	37	39	50	49
SC	Spartanburg	313	3455	8157	8.2	121	122	156	129	95	66	65	52	60	81	108	100	94
SD	Sioux Falls, Foss Fld	14944	4334	9644	9.5	158	156	215	266	190	129	94	91	121	144	212	147	161
SD	Watertown	14946	4455	9709	10.1	189	224	284	332	251	233	129	123	185	210	240	151	212
SD	Aberdeen APT	14929	4527	9826	11.2	245	234	341	413	290	244	173	177	240	249	295	202	258
SD	Huron	14936	4423	9813	10.2	164	169	229	285	214	165	131	131	165	197	239	174	187
SD	Pierre APT	24025	4423	10017	9.8	216	205	255	294	202	141	124	129	149	168	230	207	191
SD	Rapid City	24090	4403	10304	9.6	176	173	239	234	178	144	123	139	168	193	285	205	191
SD	Rapid City, Ellsworth AFB	24006	4409	10306	9.9	306	256	382	354	245	191	164	172	196	233	320	294	266
SD	Hot Springs	94013	4322	10323	8.2	90	117	155	297	219	163	94	128	126	136	179	125	152
TN	Bristol	318	3630	8221	5.9	76	105	93	85	53	40	30	30	22	37	51	59	60
TN	Knoxville APT	13891	3549	8359	7.0	133	135	156	161	85	66	53	42	47	54	99	100	94
TN	Chattanooga	13882	3502	8512	5.6	64	70	76	83	42	31	26	20	26	31	49	52	47
TN	Chattanooga	324	3503	8512	5.4	70	86	106	83	46	44	37	29	35	42	67	60	60
TN	Monteagle	126	3515	8550	5.4	77	84	84	63	32	19	18	16	19	33	60	70	48
TN	Smyrna, Sewart AFB	13827	3600	8632	5.2	76	83	87	82	42	29	22	20	22	31	58	62	51
TN	Nashville, Berry Fld	13897	3607	8641	7.4	116	114	132	117	70	67	41	33	44	57	87	80	81
TN	Memphis NAS	93839	3521	8952	6.2	84	85	93	84	54	37	25	24	29	36	67	73	57
TN	Memphis IAP	13893	3503	8959	7.9	130	137	148	125	89	57	45	42	54	61	102	110	89
TX	Brownsville, Rio Grande IAP	12919	2554	9726	10.7	231	229	281	292	277	211	185	141	102	104	150	186	199
TX	Harlington AFB	12904	2614	9740	8.8	124	175	212	207	172	160	132	129	82	75	101	109	138
TX	Kingsville NAAS	12928	2731	9749	8.5	111	130	162	183	171	157	142	115	107	77	104	98	127
TX	Corpus Christi	12924	2746	9730	10.4	189	229	260	252	199	177	158	150	107	114	157	154	178
TX	Corpus Christi NAS	12926	2742	9716	11.3	209	232	272	286	263	225	189	153	150	144	206	172	210
TX	Laredo AFB	12907	2732	9928	10.0	90	122	153	185	206	223	216	174	122	101	93	82	148
TX	Beeville NAAS	12925	2823	9740	7.3	81	101	124	129	111	85	71	61	60	52	76	72	84
TX	Victoria, Foster AFB	12912	2851	9655	7.9	134	173	198	138	112	97	67	71	53	58	103	124	109
TX	Houston	12918	2939	9517	10.1	191	216	240	258	186	139	85	74	101	117	185	159	156
TX	Houston, Ellington AFB	12906	2937	9510	6.8	86	96	114	104	81	56	38	41	56	54	82	75	72
TX	Galveston AAF	12905	2916	9451	11.0	262	261	298	257	227	200	149	144	147	148	239	217	210
TX	Pt. Arthur, Jefferson Co APT	12917	2957	9401	9.3	153	186	184	190	156	95	64	54	115	85	121	134	128
TX	Lufkin, Angelina Co APT	93987	3114	9445	6.1	65	72	76	70	45	30	27	22	26	38	56	62	49
TX	Saltillo	249	3312	9519	8.7	131	172	171	185	100	86	72	58	71	84	102	104	112
TX	San Antonio	12921	2932	9028	8.1	99	113	114	123	104	104	83	62	65	71	94	87	98
TX	San Antonio, Randolph AFB	12911	2932	9817	7.3	94	105	115	109	94	82	64	58	61	56	90	82	84
TX	San Antonio, Kelly AFB	12909	2923	9835	6.9	84	85	105	109	96	83	63	53	52	53	72	63	77
TX	San Antonio, Brooks AFB	12931	2921	9827	8.9	141	144	185	189	187	166	136	108	95	107	135	103	142
TX	Hondo AAF	12903	2920	9910	6.7	59	82	88	94	99	96	51	45	42	32	50	49	64
TX	Kerrville	12961	2959	9905	7.1	86	95	134	129	117	92	97	48	52	63	74	54	86
TX	San Marcos	12910	2953	9752	7.2	115	121	142	115	127	102	64	65	56	78	108	105	98

State	Location	Sta. No.	Lat	Long	Ave. Speed knots	J	F	M	A	M	J	J	A	S	O	N	D	Ave.
TX	Austin, Bergstrom AFB	13904	3012	9740	7.8	145	140	167	145	118	119	88	73	58	74	117	116	115
TX	Bryan	13905	3038	9628	7.0	85	104	108	103	89	74	49	48	38	47	73	89	76
TX	Killeen, Fort Hood AAF	3933	3108	9743	8.1	123	138	151	155	130	109	81	59	60	75	96	114	106
TX	Ft. Hood, Gray AAF	3902	3104	9750	9.2	163	179	198	208	162	153	121	87	72	102	147	158	146
TX	Waco, Connally AFB	13928	3138	9704	7.7	117	111	135	128	101	90	72	61	56	68	104	101	95
TX	Dallas NAS	93901	3244	9658	9.1	155	166	210	199	154	144	97	82	85	99	135	131	137
TX	Ft. Worth, Carswell AFB	13911	3246	9725	8.2	138	154	216	194	141	133	71	60	72	89	129	121	124
TX	Mineral Wells APT	93985	3247	9804	9.2	120	146	203	201	162	154	103	79	78	88	111	110	128
TX	Mineral Wells, Ft. Walters AAF	3943	3250	9803	8.7	117	135	192	184	149	136	92	71	71	82	99	103	118
TX	Santo	155	3237	9814	7.1	92	136	157	163	87	83	55	55	60	55	112	77	93
TX	Sherman, Perrin AFB	13923	3343	9640	9.1	172	164	217	213	142	121	79	73	81	106	153	153	137
TX	Gainsville	153	3340	9708	11.1	244	303	338	357	226	184	141	134	155	163	222	188	221
TX	Wichita Falls	13966	3358	9829	9.9	160	178	244	222	176	158	110	95	107	114	168	154	157
TX	Abilene, Dyers AFB	13910	3226	9951	7.7	91	104	145	147	124	102	61	51	58	66	87	87	94
TX	San Angelo, Mathis Fld	23034	3122	10030	8.9	117	154	195	184	170	147	96	86	90	90	113	108	128
TX	San Angelo, Goodfellow AFB	23017	3124	10024	8.7	108	152	191	179	169	157	85	81	91	88	114	102	126
TX	Del Rio, Laughlin AFB	22001	2922	10045	7.6	71	107	114	120	118	117	91	68	59	57	56	60	86
TX	Canadian	162	3500	10022	13.0	390	416	570	596	430	344	263	231	310	337	370	290	377
TX	Dalhart APT	93042	3601	10233	12.9	370	371	476	477	477	559	334	278	280	228	266	305	356
TX	Amarillo, English Fld	23047	3514	10142	11.7	229	279	359	329	290	240	166	135	180	200	221	222	240
TX	Childress	23007	3426	10017	10.2	152	194	272	269	221	204	118	93	117	126	128	147	170
TX	Lubbock, Reese AFB	23021	3336	10203	9.4	155	211	291	268	204	188	89	67	88	99	140	169	163
TX	Big Spring, Webb AFB	23005	3213	10131	10.1	155	197	264	266	226	216	128	101	112	124	135	139	174
TX	Midland	23023	3156	10212	8.9	91	143	146	155	133	123	94	76	85	84	89	102	109
TX	Wink, Winkler Co APT	23040	3147	10312	8.5	95	148	204	181	181	193	119	78	72	75	81	128	116
TX	Marfa APT	23022	3016	10401	7.9	128	182	165	192	146	117	84	65	85	84	88	119	125
TX	Guadalupe Pass	163	3150	10448	15.8	887	892	999	932	868	603	422	342	401	555	760	827	714
TX	El Paso	23044	3148	10629	9.8	176	257	296	299	222	173	131	111	102	128	153	163	185
TX	El Paso, Biggs AFB	23019	3150	10624	5.8	68	99	144	144	96	74	47	38	32	33	49	59	72
UT	St. George	93198	3703	11331	5.1	26	39	73	67	72	74	53	62	40	25	34	23	49
UT	Milford	475	3826	11300	10.2	179	241	228	275	302	241	220	175	167	173	172	152	214
UT	Bryce Canyon APT	23159	3742	11209	6.4	69	65	93	79	94	90	40	47	53	48	53	49	66
UT	Hanksville	23170	3822	11043	4.6	43	41	115	84	102	108	35	38	43	40	36	25	57
UT	Tooele, Dugway PG	24103	4011	11256	4.8	38	48	69	81	71	69	52	57	46	41	31	28	53
UT	Darby, Wendover AFB	24111	4043	11402	5.3	59	62	90	90	71	82	62	61	49	50	58	36	62
UT	Wendover	24193	4044	11402	5.4	48	55	91	103	84	80	57	57	43	43	47	40	62
UT	Locomotive Springs	187	4143	11255	9.3	113	129	205	193	221	215	195	202	167	125	115	93	185
UT	Ogden, Hill AFB	24101	4107	11158	8.0	102	118	127	126	130	124	124	130	122	123	99	92	119
UT	Salt Lake City	24127	4046	11158	7.7	77	84	100	100	96	96	82	106	73	68	66	71	85
UT	Coalville	174	4054	11125	3.9	26	35	39	32	30	28	17	18	37	24	17	24	28
VT	Montpelier, Barre APT	94705	4412	7234	7.2	162	167	152	118	99	97	69	60	94	103	102	125	111
VT	Burlington, Ethan Allen AB	14742	4428	7309	7.7	114	111	103	100	85	70	55	52	70	82	100	114	90
VA	Norfolk NAS	13750	3656	7618	8.8	154	182	171	139	103	84	75	80	111	127	130	128	121
VA	Oceana NAS	13769	3650	7601	7.6	135	136	150	126	84	62	52	52	83	93	97	107	98
VA	Hampton, Langley AFB	13702	3705	7622	8.5	166	187	193	166	122	86	73	81	117	137	135	146	136

Monthly Average Wind Power in the United States and Southern Canada   (Continued)

State	Location	Sta. No.	Lat	Long	Ave. Speed knots	J	F	M	A	M	J	J	A	S	O	N	D	Ave.
VA	Ft. Eustis, Felker AAF	93735	3708	7636	6.5	79	90	89	74	51	42	32	32	43	46	61	64	58
VA	South Boston	108	3641	7855	5.0	49	49	65	65	34	33	32	24	28	36	40	38	40
VA	Danville APT	13728	3634	7920	6.1	69	63	84	78	43	36	32	31	33	35	41	39	48
VA	Roanoke	13741	3719	7958	7.1	152	207	171	144	75	55	51	46	46	53	98	132	105
VA	Richmond	13740	3730	7720	6.7	59	68	79	74	49	40	34	32	39	40	47	47	51
VA	Quantico MCAS	13773	3830	7719	6.0	55	63	75	67	45	35	29	29	35	35	46	45	46
VA	Ft. Belvoir, Davison AAF	93728	3843	7711	3.8	42	60	60	42	24	15	12	12	13	19	34	44	31
WA	Spokane IAP	24157	4738	11732	7.2	92	111	106	102	73	68	55	53	58	63	80	95	79
WA	Spokane, Fairchild AFB	24114	4738	11739	7.2	118	138	134	121	92	87	67	62	74	84	92	123	100
WA	Moses Lake, Larson AFB	24110	4711	11919	6.2	68	62	98	101	79	80	56	45	61	55	53	55	68
WA	Walla Walla	24160	4606	11817	6.7	87	100	114	93	68	65	56	55	51	47	86	89	74
WA	Pasco, Tri City APT	24163	4616	11907	6.8	342	331	346	372	243	212	190	224	229	186	247	164	255
WA	North Dalles	188	4537	12109	8.0	65	72	171	189	264	279	334	284	166	93	63	64	158
WA	Yakima	24243	4634	12032	6.4	60	53	89	115	78	71	54	48	52	48	41	40	62
WA	Chehalis	792	4640	12205	6.4	109	86	82	59	53	44	45	44	43	57	80	110	66
WA	Kelso, Castle Rock	24223	4608	12254	6.9	128	107	87	65	65	46	46	39	52	66	119	133	82
WA	North Head	791	4616	12404	13.0	547	521	495	369	430	357	330	275	215	366	460	773	428
WA	Hoquium, Bowerman APT	94225	4658	12356	8.2	141	112	108	88	86	66	59	56	53	91	94	103	88
WA	Moclips	794	4715	12412	7.6	82	82	75	81	68	38	37	36	41	65	69	102	66
WA	Tatoosh IS	798	4823	12444	12.3	763	603	443	269	276	117	130	109	195	422	569	735	389
WA	Tacoma, McChord AFB	24207	4709	12229	4.6	50	48	53	49	38	30	25	23	25	31	41	41	37
WA	Ft. Lewis, Gray AAF	24201	4705	12235	3.9	36	28	30	30	23	19	17	18	17	21	23	27	24
WA	Seattle Tacoma	24233	4727	12218	9.5	194	210	211	173	130	120	94	84	104	135	147	195	141
WA	Seattle FWC	24244	4741	12216	5.6	69	61	58	48	32	28	25	24	28	44	53	69	44
WA	Everett, Paine AFB	24203	4755	12217	6.3	70	67	66	57	44	40	38	34	39	43	60	65	51
WA	Whidbey IS NAS	24255	4821	12240	7.1	185	158	148	124	75	54	43	33	50	104	160	188	108
WA	Bellingham APT	24217	4848	12232	6.3	131	132	92	66	42	43	43	35	26	56	89	119	71
WV	Charleston	13866	3822	8136	5.6	46	61	62	53	37	28	24	15	22	21	45	45	38
WV	Elkins, Randolph Co APT	13729	3853	7951	5.8	80	90	101	92	57	32	24	22	25	39	71	68	58
WV	Morgantown APT	13736	3939	7955	5.8	73	73	86	67	37	26	18	16	22	34	64	81	48
WI	Green Bay	14898	4429	8808	5.6	174	150	212	197	173	129	89	72	122	131	192	151	149
WI	Green Bay, Straubel APT	14898	4429	8808	9.3	174	150	212	197	173	129	89	72	122	131	192	151	149
WI	Milwaukee, Mitchell Fld	14839	4257	8754	10.2	198	212	246	238	195	120	95	91	129	159	229	200	175
WI	Madison, Traux Fld	14837	4308	8920	8.8	139	148	199	197	153	97	72	62	93	112	170	136	130
WI	Janesville, Rock Co APT	94854	4237	8902	7.6	158	170	203	271	234	138	108	90	112	137	212	170	167
WI	Lone Rock	143	4312	9011	7.8	117	118	125	166	116	82	60	60	75	96	133	97	103
WI	Camp Douglas, Volk Fld	94930	4356	9016	6.3	62	76	79	77	64	33	28	26	34	64	76	56	54
WI	La Crosse APT	14920	4352	9115	8.8	120	116	145	201	171	100	70	69	100	128	179	134	127
WI	Eau Claire	14991	4452	9129	8.3	96	113	102	168	155	90	83	85	100	109	134	106	112
WI	Hager City	141	4436	9232	8.4	151	132	179	221	125	98	57	77	83	125	126	119	119
WY	Cheyenne APT	24018	4109	10449	11.9	433	453	434	399	242	176	125	132	157	220	402	463	302
WY	Laramie	164	4118	10540	11.5	498	506	520	339	313	300	152	173	212	259	338	379	312
WY	Medicine Bow	165	4153	10611	12.7	773	758	825	518	343	296	250	223	328	423	544	726	490
WY	Cherokee	173	4143	10740	14.0	662	703	610	463	350	337	257	277	283	364	510	577	430
WY	Bitter Creek	172	4140	10833	12.7	477	550	623	397	297	230	150	223	210	284	370	477	364

State	Location	Sta. No.	Lat	Long	Ave. Speed knots	J	F	M	A	M	J	J	A	S	O	N	D	Ave.
WY	Rock Springs	574	4138	10915	10.6	503	476	551	357	311	264	216	189	236	250	320	394	329
WY	Granger	177	4136	10958	9.6	309	283	410	296	244	257	157	191	170	176	254	252	249
WY	Knight	573	4124	11050	10.5	278	385	375	301	261	241	175	187	276	208	233	265	254
WY	Casper AAF	24005	4255	10627	11.4	445	417	359	266	201	222	145	150	219	208	362	473	292
WY	Riverton APT	24061	4303	10827	5.4	54	52	79	75	54	49	37	31	47	38	30	40	49
WY	Sheridan	24029	4446	10658	6.6	71	71	80	94	88	73	60	56	62	65	76	64	72
WY	Cody APT	24045	4431	10901	9.3	292	303	274	342	238	189	155	176	183	218	288	303	244
NS	Yarmouth	14647	4350	6605	9.0	230	173	190	153	109	84	63	67	84	123	162	196	136
NS	Greenwood	14636	4459	6455	8.8	278	240	276	207	148	120	87	98	102	167	206	243	181
NB	Frederickton	14648	4552	6632	7.7	153	124	148	116	99	86	69	64	70	94	98	127	104
QU	Mont Jolt	14639	4836	6812	11.2	356	358	310	220	197	157	134	152	188	245	297	381	250
QU	Bagotville	94795	4820	7100	9.3	206	180	202	166	164	141	95	97	128	147	181	148	155
QU	St. Hubert	4712	4531	7325	9.3	224	213	163	153	145	138	100	83	110	136	203	177	154
ON	Ottawa	4706	4519	7540	8.2	123	120	125	111	99	79	58	53	70	88	117	100	95
ON	Trenton	4715	4407	7732	8.9	203	176	166	148	125	103	94	80	100	122	181	151	137
ON	Muskoka	4704	4458	7918	7.0	62	63	68	75	61	44	41	37	44	54	69	55	56
ON	Toronto	94791	4341	7938	8.6	198	177	157	150	110	87	69	68	85	105	177	152	128
ON	London	94805	4302	8109	9.2	239	229	233	214	139	85	66	65	82	107	173	176	151
ON	Wharton	94809	4445	8106	9.5	234	173	164	165	127	92	72	85	113	148	222	209	150
ON	North Bay	4705	4622	7925	8.5	108	121	122	121	98	85	68	67	84	90	126	104	100
ON	Sudbury	94828	4637	8048	12.2	318	392	330	324	328	290	218	195	252	294	355	312	301
ON	White River	94808	4836	8517	4.3	19	26	28	32	37	33	24	21	24	28	30	24	27
ON	Lakehead	94804	4822	8919	7.5	117	97	104	135	132	77	65	59	80	104	155	119	104
ON	Kenora	14999	4948	9422	8.6	91	95	91	111	104	74	63	70	88	96	112	86	90
MN	Winnipeg	14996	4954	9714	10.7	206	218	227	293	266	177	116	138	174	209	239	210	206
MN	Portage La Prairie	94912	4954	9816	9.5	164	162	183	217	207	122	91	107	135	169	157	157	156
MN	Rivers	25014	5001	10019	10.5	216	171	187	266	276	194	137	149	200	235	218	200	204
SA	Regina	25005	5026	10440	11.9	327	286	315	350	368	243	162	186	286	232	279	300	278
SA	Moose Jaw	25011	5023	10534	12.3	399	343	314	344	390	299	197	214	340	319	340	367	322
AL	Medicine Hat	25118	5001	11043	8.9	164	159	146	215	176	138	96	110	159	168	187	177	158
AL	Lethbridge	94108	4938	11248	12.5	625	563	356	450	370	319	202	246	279	510	546	567	419
BC	Penticton	94116	4928	11936	7.5	265	187	141	104	76	64	52	49	64	132	233	274	137
BC	Abbotsford	24288	4901	12222	5.8	135	108	89	72	46	38	32	25	28	57	86	93	67
BC	Vancouver	24287	4911	12310	6.5	72	75	85	83	53	48	52	39	48	62	80	75	64
BC	Victoria	24297	4839	12326	6.5	84	80	74	75	50	51	34	36	36	47	68	79	60

APPENDIX 2

Owning a Wind System

Installing Your Wind Turbine and Tower

You should seriously consider having your dealer install your entire WECS, or at least the tower and wind turbine. First, you will get his guarantee that the job is done right. Obviously, raising the tower and wind turbine can be very dangerous if not done properly. If you are planning to have some friends help, or are hiring help, check on your insurance situation. You will most likely find that your homeowner's policy will not cover this type of activity (Chapter 7). This appendix is included to help you understand the installation and maintenance of WECS.

The most important aspects of WECS installation that you should consider are the design of the wind turbine and tower, the two items subjected to wind loads. Good design, however, is not enough. These units must be properly installed. This includes appropriate grounding for lightning strike protection. After installation, maintenance must be performed as required to assure continued reliable service. Each of these items, if performed properly, will contribute to the ultimate safety and efficiency of your wind system.

Figure A2-1 diagrams the step-by-step sequence of WECS ownership. If any block in this diagram is omitted, a potential ownership problem is created.

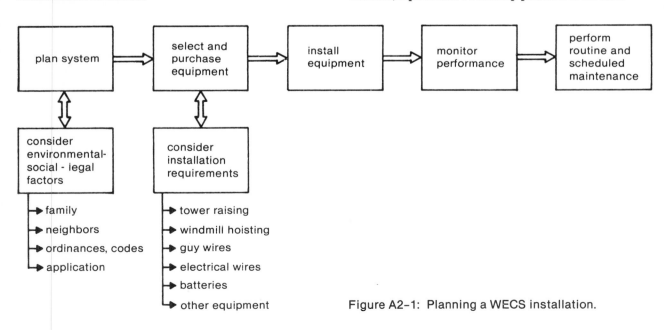

Figure A2-1: Planning a WECS installation.

Consider the neighborhood resident who hoists aloft a wind turbine without considering his neighbors' feelings; something like the problems which arose early in the history of television antennas. This situation is a little like "I don't have one, so why should you?", or "That's an ugly machine, can't you hide it over behind that tree?", or "That's a very noisy propeller, isn't it?" We have heard these comments before; some are legitimate, others are not.

In at least one United States protectorate, it would be illegal to have any form of auxiliary power source, wind included, if the utility mains exist at the edge of your property. This is not the case in the United States, but local building ordinances and codes may prohibit installation of towers tall enough to make wind power practical, or they may require a tower designed to withstand loads so high that the tower cost makes the entire system economically impractical.

Another possibility is that the entire proposed system meets all requirements but cannot be installed for lack of space to install the equipment. For example, perhaps the tower cannot be raised within the confines of the area, or there are too many tall trees. We mention these aspects of system planning even though they only rarely apply to specific installations.

Again we must emphasize the possible dangers involved in raising a tower and a wind turbine. If you haven't had experience with this sort of thing and insist on doing your own, get a knowledgeable friend to go over your plans and be there for the most critical jobs.

Tower Raising

To raise a tower, you can either assemble it on the ground and tilt it up (Figure A2-2) or assemble it standing up. The first method requires assembly of all components, guy wires, and as much of the wind turbine equipment as possible on the ground. The base of the tower is then fixed to a pivot to prevent the tower

from sliding along the ground and a rope is tied from the tower, over a *gin pole* (Figure A2-2) to a car or winch. Moving the car pulls the tower up. The gin pole serves in the initial stages to improve the angle at which the rope pulls on the tower.

In the case of the tower being pulled up by a rope tied to a car bumper, it might be well to pull with the car backing up so the driver maintains a clear view of the action. Also, an effective, foolproof communications link must be established and maintained between the driver and the person who is directing the operation. If not, towers pulled over center, bent, broken cables, and a host of other crises are likely to beset the tower crew.

Tower-raising techniques such as this are usually described in the owner's manual or installation instructions that come with the tower. Since each tower has a different load rating and different installation requirements, it is not possible to discuss the details of tower raising.

Many towers, such as a freestanding octahedron module tower, can be erected in place. This is usually done by assembling the first few bays on the ground, standing these up, then assembling the remaining bays while standing on each successive lower bay.

Wind Turbine Raising

To raise a wind turbine, you can: hoist a completely assembled machine up an already erected tower (Figure A2-4); hoist a partially assembled machine up an already erected tower, completing the assembly aloft; or tilt the tower up with the wind turbine already installed. The first two methods rate the title "traditional"; the last often is not possible or safe.

Personal experience will tell, but generally the amount of enthusiasm one has for doing anything atop a tower decreases rapidly with increasing tower height. This serves as a token justification for ground

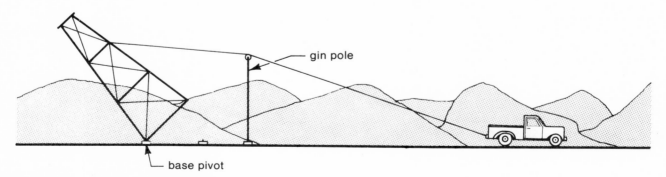

Figure A2-2: Tower raising with a "gin pole."

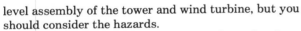
Figure A2-3: Hoisting a wind machine up the tower.

level assembly of the tower and wind turbine, but you should consider the hazards.

Tilting up a tower with the wind machine already installed imposes additional loads on the tower. The compressive load at the base pivot and the bending loads where the rope is attached will be much greater. Most likely, you will have to provide extra bracing, but in any case you should consult the manufacturer about these loads or rely on a competent installation crew.

Consider also that if anything goes wrong, you stand to lose the entire tower and wind machine. Risks go up rapidly as tower height and wind machine weight increase. This is not to say, however, that this method will not work; it does, but the individual installation will dictate the method. For sturdy towers of 20 to 40 feet, experience has shown that tilting the whole works aloft can work.

For taller towers, you can expect to hoist the wind turbine up an already erected tower. This will require a block and tackle supported aloft and an extra rope to the wind machine. The block and tackle is used to lift, while the rope is tugged at from the ground to keep the wind machine from banging into the tower as it journeys upward.

Consider that with a hand-operated block and tackle, you can feel what is happening. Tail vanes

snagged on a guy wire may not be detected until damage has occurred if you use a winch or auto-pulled hoist. The support structure that holds the block and tackle to the tower top must not bend or otherwise yield to the loads of the wind turbine. You can test it by hoisting up the wind machine with a volunteer adding extra weight. Remember that you or another person must depend on this hoist to suspend the machine over its mount while you bolt it down, maybe 60 or 80 feet in the air. This is something like changing engines in a Volkswagen that is hanging much higher than you would care to fall.

Points to Remember

Points to remember while doing any wind turbine installation:

1. Hard hats are required, as tools, bolts, and other objects seem to be routinely dropped from aloft.
2. Climbing safety belts must always be used.
3. Make provision for preventing the wind turbine from operating until it is fully installed. Feather the blades, tie them with a rope, or otherwise lock them. One of Murphy's laws says that whatever can go wrong, will. While you may not detect a breeze on the ground, there may be

156

enough wind aloft to create an unpleasant surprise about halfway into the installation process.

4. Perform installations when no wind is expected, and start early. Installations always take longer than expected; bolting a wind turbine aloft after dark is to be avoided.

5. Plan and practice the entire installation process very carefully. The process should cover such details as who has which bolt in which pocket, when said bolt is to be installed with what tool, and by whom.

6. Create an alternative plan. This plan is designed to be used as a contingency if something goes wrong (e.g., the main bolt gets dropped and the wind is coming up).

7. Tools and parts can best be carried aloft in a carpenter's tool belt, available at most hardware stores.

8. Gloves and a warm jacket with lots of pockets will be useful in keeping you warm and able to work and finish the installation, even if a wind starts to come up; the pockets will save tiring trips up and down the tower.

9. Pay close attention to the strength of ropes, pulleys, or other auxiliary equipment you may use in hoisting equipment aloft. For example, if you use standard 7/16-inch climbing rope for hoisting, a nylon rope will withstand about 3900 pounds (wet strength), while a manila rope is rated at 2600 pounds. If you tie a knot in the rope, you will reduce its strength to about 60 percent of the original, and if you pull it around a tight radius—like a bolt—you reduce the strength of the rope to 80 percent of its rating. Naturally, smaller ropes have lower load ratings. If you are hoisting aloft a 400-pound machine, and you want a factor of safety of about 4 (a minimum you should plan for), you are going to need a rope and other equipment capable of hoisting 4 × 400 = 1600 pounds. If you have a manila rope rated at 2600 pounds and you tie a knot at its attachment, the rope is really good for 0.6 × 2600 = 1560 pounds. This rope is the minimum strength to consider for the job.

Wiring

Wire size, wire routing, and lightning protection are important considerations. Wire size is determined by the current (amps) that will flow and the length of wire. In general, you select wire sizes to limit the line voltage loss to a small percentage (Figure A2–4). Using Figure A2–4 and the following simple equation, you can calculate the wire size you need.

157

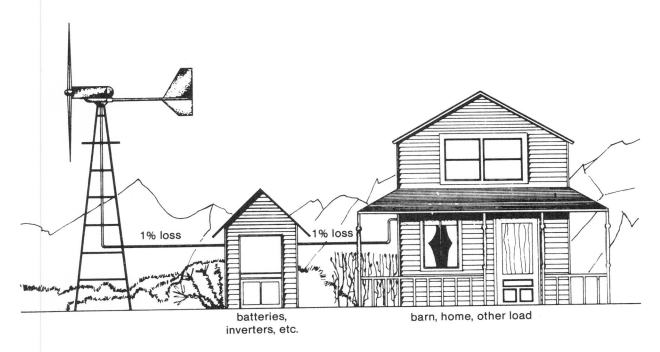

Figure A2–4: Typical line voltage loss allowance.

For aluminum wire: Circular area size = 35 × amps × feet of length/volts line loss

For copper wire: Circular area size = 22 × amps × feet of length/volts line loss

Circular area size is converted to wire gauge size from the following table:

Wire Gauge (AWG)	Circular Area Size (Circular Mills)
14	4,017
12	6,530
10	10,380
8	16,510
6	26,250
4	41,740
3	52,640
2	66,370
1	83,690
1/0	105,500
2/0	133,100
3/0	167,800

For example, calculate the wire size required for a copper wire run of 25 feet, carrying 25 amps at 24 volts, with a 1 percent line voltage drop.

Solution: from Figure A2–4, 1% = 0.24 volt

then,

Circular area size = 22 × 25 × 25 / 0.24 = 57,291

From the table, 57,291 is between wire gauge 2 and 3. Select wire gauge 2 for conservative selection.

Wire routing, except for any deviations necessary for lightning protection, is a matter of direct routing, adequate support to prevent wind or mechanical damage to the wind turbine or wire insulator and adequate separation for prevention of electrical short circuits. Wires that are routed down the tower should be tied to the tower every few feet or fed through a conduit to preclude wind damage.

Lightning Protection

The secret of protecting wind power equipment is to install a good ground wire. This means that you must electrically tie the tower if it is metal, or the wind turbine itself (if it is mounted on a wooden tower) to the earth. All tower guy wires must be grounded by methods discussed here, and electrically connected to each other as well as the tower.

The National Electrical Code (available at libraries) specifies ways in which various towers and antennas are to be grounded, and the information is useful for WECS. In general, an underground metal water pipe is desirable to use as the ground. In its absence, one or several (use several in dry ground) heavily galvanized pipes (1½ inch diameter is adequate, 1¼ inch minimum), or a ½-inch copper rod, are driven into the ground to a minimum depth of 8 to 10 feet. Instead of rods, sheets of copper-clad steel or galvanized iron about 3- by 3-feet in dimension can be buried about 8 feet deep horizontally and connected to each other as well as to the tower. This forms an electrical "ground plane." When more than one rod is used, all should be electrically connected to each other as well as to the tower. Wire size for grounding should not be smaller than number 10 copper wire (usually number 6 or bigger), and it is usually a bare wire.

Electrical wires can be protected by a spark arrester (Figure A2-5). These are available at electrical supply houses. In the case of a ground wire on a wooden tower, this wire will protect both the wind turbine and the electrical wires. You should consider adding a ground wire up the entire height of a metal tower as

Figure A2-6: Battery bank installation.

corrosion eventually weakens the ground connection of these towers. This ground wire should have a cross-sectional area at least as great as the total of the two wires it is protecting.

Installing Other Equipment

A wind system that generates electricity to be stored in batteries is a good deal more complex in its installation requirements than, say, a farm-type water pumper. Provision must be made to install batteries, inverters, controls, wires, and perhaps other equipment. In all cases, follow the manufacturer's recommendations, but here are a few items to consider: It is generally desirable to install batteries near the wind generator, especially if lower voltage is to be inverted up to higher voltage. Higher voltage means lower current for any given load. This means wires can be smaller in size, and line loss is reduced.

Batteries should be installed in a cool (but not cold), dry, well-ventilated space and should be well insulated to prevent large temperature changes (Figure A2-6). Some installations have the batteries in a small lean-to built alongside a home or barn; others have the batteries in a basement.* Basement installation is reasonable, but gas formed by the batteries produces an

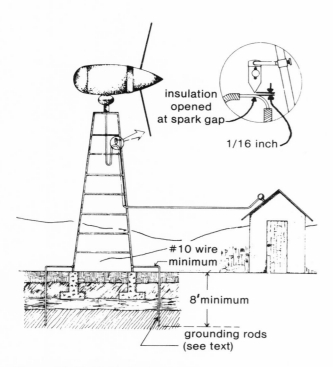

Figure A2-5: Lightning protection.

---

*Electrical equipment, and especially large battery banks, should be protected by a locked door from vandals and small children. A large wrench dropped across large battery terminals could result in an explosion or a fire.

explosion hazard if there are open flames or sparks in the same area. Special caps can be purchased for lead-acid batteries that reduce gassing by catalytic conversion of the hydrogen gas back into water. Some batteries are even offered without vents. You must know the gassing characteristic of your battery before selecting an installation site. Inverters of the motor-generator type should be bolted to a bench or mounting pad. These units don't vibrate much, but their bearings last longer with a solid installation. Static (electronic) inverters generate heat, which means ventilation is the prime installation requirement, as is a dry space free from heavy dust exposure. Most electrical controls and load monitor equipment may be installed in the same space as inverters and batteries.

Maintenance

From a historical standpoint, you might purchase a good automobile and drive it an average of 50 mph for 100,000 miles. This translates to 2000 hours of operation. You likely would have the car serviced every 5000 miles, or 100 hours. Sales brochures for small WECS, on the other hand, sometimes speak of 20 years of trouble-free operation. Factory representatives talk of customers asking how long their wind turbine will last before it needs fixing.

There are 8760 hours in a year. If your wind turbine operates just one-fourth of the hours in one year—a reasonable number—it will have as many hours on it in one year as your automobile does when you trade it in. It is also reasonable to expect to change the oil, grease a bearing, or change the brushes just once a year to get long-life performance out of a machine. Here are some of the factors you should consider before making the final selection of your system:

- Maintenance history of WECS components
- Can routine maintenance be performed easily up on a tower, or must the machine be lowered?
- Frequency of expected routine maintenance
- Nature of monthly, or yearly, expected routine maintenance: lubrication, component replacement, and inspection
- Number of different tools required to perform maintenance tasks
- Availability and cost of spare parts
- Completeness of owner's manual/maintenance documentation
- Relative safety: can machine be shut off, is there sufficient blade clearance from maintenance personnel, and are there exposed shafts, wires, and potential hazards?
- Is a factory-trained, experienced installation/maintenance organization available?

Answers to many of these questions are available directly from dealers and other users. Some questions will never have answers but are subject to your best estimate during product evaluation.

The air mass flowing past your wind machine is full of dust and grit, which, over a long period of time, gets into the various components, including bearings, transmissions, and generators. Changing the oil, greasing the bearings, inspecting generators or pumps is the way in which you or your mechanic can monitor the system and prevent rapid wear from such environmental conditions. Evidence of rust, loose wires, worn bushings, and so on should be on an inspection list, which is used each time the wind turbine is inspected.

Blades or vanes which are exposed to the wind are subject to impact from hail, rain, ice, and rocks. Any inspection should include examination of these blades. Wooden blades might need fresh paint; fiberglass and metal blades might also need similar service.

Vibrations in the wind turbine can cause bolts and nuts to loosen and parts to fatigue and fail, wires to break, and so on. This is another area for thorough examination. Properly bolted joints will not fatigue. These are items that should be inspected as a routine, preventative procedure. Once a year is a usual interval for performing this type of maintenance. You or your mechanic would normally schedule this work for a nonwindy day. Most of the items listed above rarely, if ever, require any maintenance action but should be inspected anyway. Some owners climb their tower once a month, just to see that everything is in order. In any event, manufacturers usually have a recommended inspection routine.

Maintenance should be scheduled following any extreme wind or hailstorm. These conditions warrant a brief inspection.

Electrical equipment such as batteries and inverters require cleaning, water checks, and terminal inspection. Water pumps need to be checked for leaks. The list seems endless, but each check is necessary. Time spent in inspection, cleaning, and lubricating will be returned in extended service life.

Environmental Impact

Along with all other factors, planning a WECS installation involves consideration of its impact on the environment.

Small systems are not suspected of producing harmful effects, based on almost a century of experience with hundreds of thousands of wind machines.

Studies are also being conducted to test the impact of wind turbine rotors on TV reception. Again, small systems are not really suspect here, unless one installs

Figure A2-7: A series of small wind systems at an exposition.

a dozen or so of them, in which case all of these units collectively might affect electromagnetic waves. These effects would be highly local in nature.

Typically, wind turbines are installed far enough from dwellings that ambient wind noise is higher than machine or rotor noise. You should keep this characteristic in mind and determine for yourself the noise characteristics of the wind machine you plan to use. Noise comes from blade tips, transmissions, bearings, and generators. Some machines are noisy; others are not.

Towers with guy wires usually require care to preclude guy wires from encroaching upon existing or planned easements. Tower footings may extend deep into the ground. It is usually unacceptable to install a tower directly over a septic tank or water main, but it has been done.

The visual impact of a wind system is an area for personal taste—not just your taste but that of your neighbors. No words of caution written here will substitute for your own investigation into the potential reaction to your planned wind system.

All of the notes in this Appendix have been derived from the collective experience of the authors and from interviews with respected wind energy technicians and consultants. While many of the points raised here were written as warnings, the frequency of occurrence is low for any problem area mentioned. We feel that such occurrences will remain low and owners of a small WECS will enjoy years of satisfactory service from their machines if careful consideration is given to these and other factors related to WECS ownership.

English-Metric Units Conversion Table

Physical Quantity	This In "English" Units	Equals, in Metric* Spelled out	Symbolic	Reciprocal†
DISTANCE	1 inch	2.54 centimeter	2.54 cm	0.3937
	1 foot	0.3048 meter	0.3048 m	3.281
	1 yard	0.9144 meter	0.9144 m	1.094
	1 mile	1.609 kilometer	1.609 km	0.6215
AREA	1 square inch	6.452 square centimeter	6.452 cm²	0.155
	1 square foot	0.0929 square meter or	0.0929 m²	10.76
		929 square centimeters	929 cm²	0.001076
	1 square yard	0.836 square meter	0.836 m²	1.196
	1 acre	4,047 square meters or	4.047 m²	0.000247
		0.4047 hectare	0.4047 h	2.47
	1 square mile	2.590 square kilometers or 259.0 hectares	259.0 h	0.00386
VOLUME	1 cubic inch	16.39 cubic centimeters	16.39 cm³	0.0610
	1 pint (liquid)	473.2 cubic centimeters	473.2 cm³	0.002113
	1 quart	946.4 cubic centimeters	946.4 cm³	0.001057
		or 0.9464 liter	0.946 l	1.057
	1 gallon	3.785 liters	3.785 l	0.2642
	1 cubic foot	0.0283 cubic meter	0.283 m³	35.3
	1 cubic yard	0.765 cubic meter	0.765 m³	1.308
	1 acre-foot	0.1233 hectare-meter	0.1233 h m	8.11
VELOCITY	1 foot per hour, minute or second	0.3048 meter/hour, minute, or second		3.281
	1 mile per hour	0.4470 meter per second	0.4470 m/s	2.237
	1 knot	0.5145 meter per second	0.5145 m/s	1.944

*Multiply quantity known in British units by this number to get metric equivalent.
†Multiply quantity known in metric units by this number to get British equivalent.

Physical Quantity	This In "English" Units	Equals, in Metric* Spelled out	Symbolic	Reciprocal†
ENERGY (OR WORK)	1 watt-second	1.000 joule = 1.000 newton-meter	1.000 J	1.000
	1 foot-pound	1.356 joule	1.356 J	0.7375
	1 Btu	1.055 kilojoule	1.055 kJ	0.948
	1 watt-hour	3.60 kilojoules	3.60 kJ	0.2778
	1 horsepower-hour	2.684 megajoules	2.684 MJ	0.3726
	1 kilowatt-hour	3.60 megajoules	3.60 MJ	0.2778
POWER	1 horsepower	745.7 watts or 0.7457 kilo-watt	745.7 W 0.7457 kW	0.00134 1.341
	1 joule per second	1.000 watt	1.000 W	1.000
	1 Btu per hour	0.293 joule per second	0.293 J/s	3.41
TEMPERATURE	1 degree Fahrenheit	5/9 degree Celsius (Centigrade) for each Fahrenheit degree above or below 32°F	$5/9 \times (T_F - 32)°C$	1.8 degree Fahrenheit for each Celsius degree plus 3
SPECIAL COMPOUND UNITS	1 Btu per cubic foot	37.30 joules per liter	37.30 J/l	0.0268
	1 Btu per pound of mass	2.328 joules per gram	2.328 J/g	0.4296
	1 Btu per square foot per hour	3.158 joules per square meter	3.158 J/m²	0.3167
	1000 gallons per acre	0.0935 centimeters depth		10.70
	1 pound of mass per cubic foot	16.02 grams per liter	16.02 g/l	0.0624
MASS	1 ounce	28.35 grams	28.35 g	0.03527
	1 pound	453.6 grams or 0.4536 kilogram	453.6 g 0.4536 kg	0.002205 2.205
	1 ton (short, 2000 pounds)	0.907 megagram or 0.907 metric ton or 0.907 tonne	0.907 Mg 0.907 t 0.907 t	1.102 1.102 1.102
TORQUE	1 inch pound	0.1130 meter-newton	0.1130 m-N	8.851
PRESSURE	1 pound per square foot	47.88 newtons per square meter	47.88 N/m²	0.02089
	1 pound per square inch	6.895 kilonewtons per square meter	6.895 kN/m²	0.11240
	1 millimeter of mercury	133.3 newtons per square meter	133.3 N/m²	0.0075
	1 foot of water	2.989 kilonewtons per square meter	2.989 kN/m²	0.3346
	1 atmosphere	0.1013 meganewton per square meter	0.1013 MN/m²	9.87

Continued on page 164.

Physical Quantity	This In "English" Units	Equals, in Metric* Spelled out	Symbolic	Reciprocal†
FLOW	1 gallon per day	0.04381 milliliters per second	0.04381 ml/s	22.824
	1 gallon per minute	63.08 millileter per second	63.08 ml/s	0.01585
	1 cubic foot per minute	0.4719 liter per second	0.4719 1/s	2.119
	1 cubic foot per second	28.32 liters per second	28.32 l/s	0.0353
FORCE	1 ounce	0.2780 newton	0.2780 N	3.597
	1 pound	4.448 newtons	4.448 N	0.2248
	1 ton (2000 pounds)	8.897 kilonewtons	8.897 kN	0.11240

APPENDIX 4

Additional Wind Behavior and
Site Selection Information

The purpose of this appendix is to present some more technical material that adds to the information on wind behavior and site selection given in Chapter 3.

Winds Variations with Time, Including Maximum Storm Winds

A typical hourly wind speed record for eight days is shown in Figure A4-1 (the average annual wind speed for this site is 7 mph). These data points are connected with straight lines to make the data easier to follow, though the wind speed actually varied greatly during each hour. The dashed lines connect the resulting data points of wind power (watts per square meter) generated for a typical windmill with an overall efficiency of 30 percent and cut-in and rated speeds* of 8 mph and 25 mph, respectively. For clarity, the horizontal line for zero wind power has been raised above the zero wind speed line. This figure emphasizes how much the wind and windmill power change with time and shows why a site survey cannot be accomplished in a few hours.

Maximum Storm Winds. What are the chances your windmill will get blown down during its lifetime? The maximum wind speed you will experience is a problem involving chances or probability—there is no definite answer. As a start, Figure A4-2 gives the average annual maximum fastest mile at 50 feet above the ground at selected stations in the United States. Also listed are more useful numbers, the fastest miles with a 2½ percent and a 10 percent chance of oc-

curring in the next 25 years. By *fastest mile* is meant the speed indicated by the passage of one mile of wind in the shortest time. For instance, if an anemometer experienced a steady 60 mph wind, in one minute one mile of wind would pass; or if it were a 120 mph wind, a mile of wind would pass in half a minute. Twenty-five years is probably a reasonable life for a well-maintained wind generator, if it doesn't blow down, and a 10 percent chance of that happening is probably reasonably small odds for most situations. The difference between the first two columns in the accompanying table gives an indication of the intensity of the storms in the area. For instance, the large differences in these two wind speeds for Miami, Wilmington, and Hatteras is caused by the occasional violent hurricane that passes close by.

Of course, it doesn't usually take a half minute to a minute of high wind to knock down a structure if that wind speed will do the job. Just a gust to that speed would probably knock it over. A very approximate value for the maximum gust speed is about 30 percent greater than the fastest mile. This means that the maximum gust with a 10 percent chance of occurring in 25 years ranges between about 75 and 165 mph for the locations listed in Figure A4-2. Remember, it was shown in Chapter 2 that wind pressure increases with the square of the velocity.

When selecting your windmill and tower you should be concerned with the survival speeds listed. Your county agent, a structural engineer or meteorologist may be able to supply maximum wind speeds for your general area. The Energy Research and Development Administration (ERDA) test center for windmills the size you will be interested in is located in Rocky Flats,

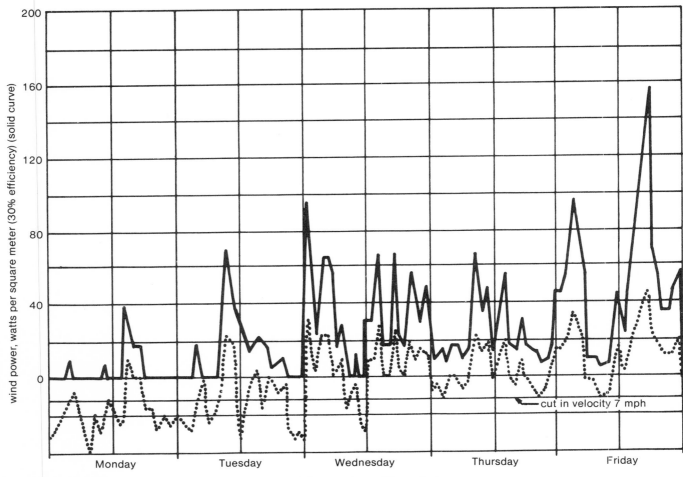

Figure A4-1: Eight-day sample of hourly wind speed data and resulting wind power.

Station	Average Maximum Speed (mph)	2½% Risk in 25 Years (mph)	10% Risk in 25 Years (mph)	Station	Average Maximum Speed (mph)	2½% Risk in 25 Years (mph)	10% Risk in 25 Years (mph)
Tampa, Fla. (A)	52	95	86	Omaha, Nebraska (A)	59	124	109
Miama, Fla.	54	143	123	El Paso, Tex. (A)	58	80	75
Wilmington, N. C. (A)	67	146	128	Albuquerque, N. M. (A)	61	112	100
Hatteras, N. C.	62	129	113	Tucson, Ariz.	50	85	77
Dallas, Tex. (A)	52	84	77	San Diego, Calif.	36	66	59
Washington, D.C. (A)	50	92	83	Cheyenne, Wyo.	63	97	90
Dayton, Ohio (A)	60	103	93	Rapid City, S. D.	66	99	92
Atlanta, Ga. (A)	50	87	78	Bismarch, N. D.	66	92	86
Abilene, Tex. (A)	63	131	115	Great Falls, Mont.	65	82	78
Columbia, Mo. (A)	56	87	79	Portland, Ore.	57	91	83
Kansas City, Mo.	55	90	82	New York, N. Y.	58	82	76
Buffalo, N. Y. (A)	58	99	90	Pittsburgh, Pa.	52	83	76
Albany, N. Y. (A)	52	94	84	Fairbanks, AL.	37	78	69
Boston, Mass. (A)	59	119	73	Nome, AL.	61	106	96
Chicago, Ill.	51	79	73	Elmendorf AFB, AL.	45	81	72
Cleveland, Ohio (A)	59	88	81	Shemya Island, AL.	70	101	94
Detroit, Mich. (A)	49	77	71	Hickam AFB, HI.	45	86	77
Minneapolis, Minn. (A)	52	107	95				

Figure A4-2: Extreme annual wind speed (fastest mile) at 50 feet above the ground at the given stations; (A) denotes airport station.

166

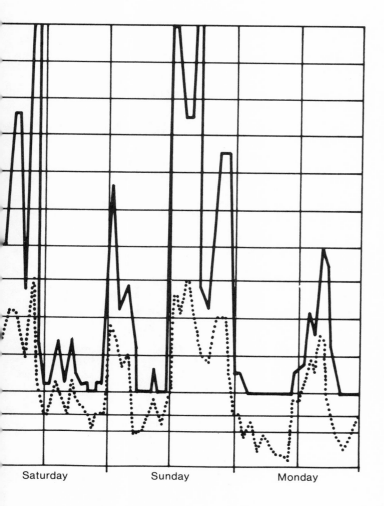

Saturday	Sunday	Monday

centage of time the wind blows in that direction (toward the center of the rose). Each circular arc represents 5 percent, so a bar that extends from the bulls-eye just to the second arc indicates that the wind blows in that direction for 10 percent of the time.

In large, flat areas of the country, the shape of the wind rose should change slowly with distance. Near large bodies of water or hills of any sort the changes can be great in short distances. You can construct the wind roses for weather stations close to you by simply obtaining a copy of their "Percentage Frequency of Wind Direction and Speed" table. Figure 3-22 shows an example. The second column from the right has the percentages needed for the plot. Note that the length of the lines in a wind rose plot does not indicate the wind speed. If the maximum wind speed has the same direction as the maximum percentage, up to half the wind energy can come from just one of the 16 directions. If you must have some blockage effect, from a silo, for instance, you may want to construct the wind roses for your most critical months for wind energy usage. This could help you avoid selecting a poor site.

Wind Power Variations with Height in Flat Terrain

In Chapter 3 wind speed profiles were described. We will expand on this subject here. In flat terrain with no land-sea or valley-mountain effects and with a uniform ground cover, the shape of the wind speed profile depends only on the roughness of the ground cover and how the air temperature changes with height. If the shape of the wind speed profile is known and the speed at any height is known, the speed at any other height can be determined. At the wind speeds that appreciable wind power can be produced, the temperature effects are nearly always unimportant. Therefore, for a large area of flat, uniform country with a good wind speed, a very important conclusion can be drawn:

The shape of the wind speed profile depends *only* on the *roughness* of the land (or water) and *not* upon the actual *wind speed.*

"So what?" you ask. Well, if you live in a flat area and the terrain is quite uniform (e.g., fields with scattered buildings and trees), you can estimate the usual shape of your wind speed profile using Figure A4-4. Then, if you determine your average wind speed at one height, you can use that profile to estimate the speed at the height you wish to locate your windmill.

Each of the six sets of curves in Figure A4-4 consists of a wind speed profile shape and the cor-

Colorado.* This area experiences very severe winds. If you believe you may have a severe wind storm problem you or your engineer may wish to consult with them.

Wind Roses for the United States

Figure 3-10 of Chapter 3 shows very generally the average wind speeds in the continental United States. Since the local wind speed can change a lot in a few hundred feet, this curve is of limited use. The monthly wind power listings in Appendix 1 provide data useful to most potential windmill owners. It does not have, however, any data on wind directions.

Figure A4-3 presents annual wind roses for a number of sites in the United States. The number in each bulls-eye is the percentage time there is no wind blowing. As in Figure 3-5, the wind rose example in Chapter 3, the length of each bar represents the per-

---

*Rockwell International, Atomics International Division, Rocky Flats Plant, P.O. Box 464, Golden, CO 80401.

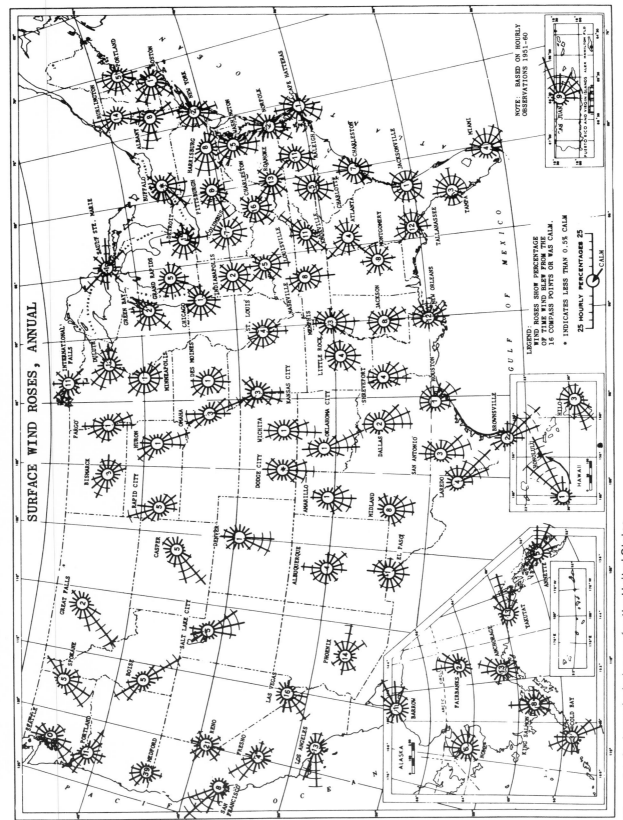

Figure A4–3: Annual wind roses for the United States.

responding wind power profile (the cube of the wind speed).* Compare your terrain with the descriptions in these figures and select the one closest to your situation. If you are technically inclined, you can even do your own interpolating to obtain a closer estimated profile. Since a good minimum height for placing a wind anemometer is 30 feet if it is well clear of obstructions (see the last section of this chapter), all the wind profile curves pass through 1.0 at 30 feet. What is plotted is the ratio of wind speed and power at other heights to the speed and power at 30 feet.

The curves in Figure A4-4 are to be used in the following manner. Suppose you find that your average wind speed is 10 mph at 30 feet, the surrounding country is nearly flat, and fits the description for part B – "4-inch-high grass." What is the average wind speed at 70 feet? At 100 feet? From the wind speed curve at 70 feet, we read 1.13, and at 100 feet, 1.19; therefore,

$$\text{wind speed at } 70 \text{ feet} = 1.13 \times 10 = 11.3$$
$$\text{wind speed at } 100 \text{ feet} = 1.19 \times 10 = 11.9$$

How much more power is available at 70 and 100 feet than at 30 feet? From the wind power curve at 70 feet we read 1.43, and at 100 feet, 1.66, so:

$$\text{additional power available at } 70 \text{ feet}$$
$$= (1.43 - 1) \times 100 = 43\%$$

$$\text{additional power avilable at } 100 \text{ feet}$$
$$= (1.66 - 1) \times 100 = 66\%$$

Notice that the descriptions in Figure A4-4 go from the smoothest possible earth surface to a very rough surface ("woodland" for part F). As roughness increases, the wind speed increases more rapidly with height.

Actually, the zero height for these profiles should be taken as close to the top of the densely covered part of the ground cover. For the plots in parts A, B, and C this effect is small, so there is a negligible effect for just using the height from ground level. However, there is a considerable displacement effect for parts E and F. Unfortunately, no standard rule can be given for the ground height to use at the bottom of these curves. It is probably best to use the height where the branches from adjoining trees no longer overlap, where the tops of the individual trees are separated.

The advantage of mounting your windmill at a

*A note for engineers and scientists: these wind speed profiles are described by $u/u_{30} = (z/30)^a$, where $a = 0.11, 0.14, 0.16, 0.19, 0.23,$ and 0.28. The power profiles use exponents three times these values. The ground cover descriptions include the authors' interpolations from available data.

greater height to obtain more power is rapidly offset by the increasing tower costs with height. However, the rougher the surface, the more rapidly the power will increase with height. Consider a woodland with a few trees as tall as 60 feet but with branches essentially covering all the space between the trees up to 40 feet. We use the profiles in Figure A4-4, part F – but note that the profile height should be taken as starting from about 40 feet. How much more power is available at tower height of 100 and 130 feet compared to 70 feet? We read the power profile curve at

$$70 - 40 = 30 \text{ feet: } 1.00$$
$$100 - 40 = 60 \text{ feet: } 1.79$$
$$130 - 40 = 90 \text{ feet: } 2.52$$

The additional power available at 100 feet = (1.79 − 1) × 100 = 79 percent, and the additional power available at 130 feet = (2.52 − 1) × 100 = 152 percent. In a case like this, the cost of the considerable additional tower height may be worthwhile.

The possible errors using the lower curves in Figure A4-4 can be large. In rough terrain, wind speed measurements should be made closer to the anticipated windmill height to reduce such errors. No effort has been made to present appropriate curves for the flow over very rough forests or cities, as unacceptable errors can easily result in their application.

The profiles in parts D and E must be used with care close to trees. The effect of individual trees and wind breaks is described later.

Wind Over Changing Terrain

Often a possible windmill location will be in flat terrain, close to the boundary of two types of ground cover. At the start of a new ground roughness, only the wind relatively close to the ground is affected. The further downwind from the change, the higher the wind profile will be affected by the new cover. How the profile changes downstream of a changing roughness is indicated in the sketches of Figure A4-5A for a change from smooth to rough terrain, and in Figure A4-5B for a change from rough to smooth terrain. The wind, after blowing across some of the new terrain, will have the lower part of its wind speed profile with a shape corresponding to the profile in Figure A4-4 for the downstream roughness. Above a thin transition region, the upper part of the profile will correspond to the wind profile for the roughness before the change in cover. Figure A4-6 consists of five curves that give the transition height between the various profiles of Figure A4-4.

As an example using these figures, suppose a farmer

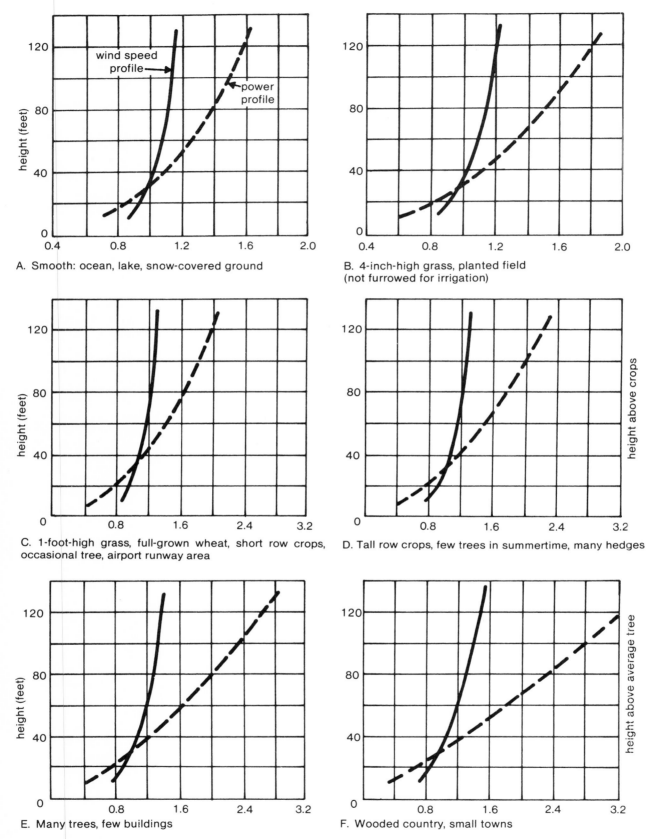

A. Smooth: ocean, lake, snow-covered ground

B. 4-inch-high grass, planted field (not furrowed for irrigation)

C. 1-foot-high grass, full-grown wheat, short row crops, occasional tree, airport runway area

D. Tall row crops, few trees in summertime, many hedges

E. Many trees, few buildings

F. Wooded country, small towns

Figure A4–4: Wind speed profiles and wind power profiles for various types of flat terrain.

170

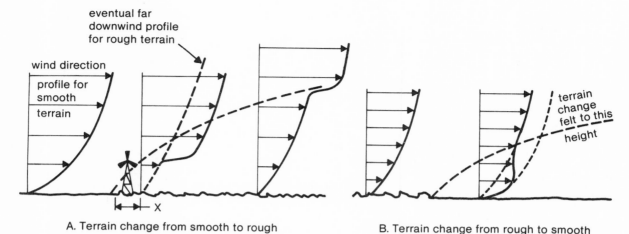

A. Terrain change from smooth to rough   B. Terrain change from rough to smooth

Figure A4-5: Wind speed profile near a change in terrain.

in flat country has a windmill mounted 65 feet high in a large pasture and conditions describing part B of Figure A4-4 appear to be good for his terrain. His average monthly energy output has been 1000 kilowatt-hours. A neighbor wishes to erect the same windmill near his house but is surrounded by 300 feet of orchard trees 20 feet high. How high should he mount his windmill to obtain the same output, assuming the orchard is the only obstacle causing the wind to be different?

The appropriate ground roughness for an orchard is

represented by conditions for part F in Figure A4-4. The transition, then, is from B to F, and the key to Figure A4-6 indicates that curve 4 should be used. At 300 feet, the distance from the edge of the orchard to the windmill, the transition from the F profile to the B profile will be about 60 feet above the trees, so about 80 feet above the ground. His wind, then, at 65 feet will be reduced by the orchard, and he should go at least 30 feet higher, allowing some margin for uncertainties, to clear the transition region and be sure to obtain at least the same monthly energy output.

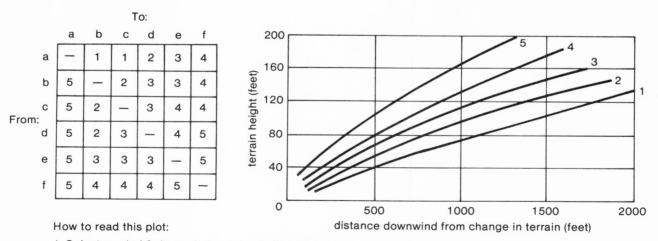

How to read this plot:

1. Select upwind & downwind terrains in Fig. A4-4.

2. Enter table above with appropriate letters, select number.

3. Use curve with that number.

Figure A4-6: Transition height in wind speed profile.

Wind Over a Ridge

The effects of wind blowing across a ridge consisting of flat sides elevated to form a triangular cross section have been investigated in a special wind tunnel. This wind tunnel can reproduce, on a small scale, the atmospheric wind speed profile. By using scale models in a wind tunnel, a steady, reproducible wind is created. Even more important, wind speed measurements can be made that will closely represent those of the flow over an object in nature that has been scaled down in size. To make the equivalent wind speed measurements in nature would be very expensive.

Results of these measurements over ridges can be very useful if you live on or near a ridge. Figure A4-7 shows the measurements of wind speed profiles at different locations approaching and on the windward side of three different ridges. The ridges have sides with constant slopes of 16.7, 25, and 50 percent. (16.7 percent means 16.7 feet of vertical rise in every 100 feet of horizontal distance; by comparison, 25 percent is the slope of a typically steep driveway, and highways have slopes considerably less than 10 percent.)

The approaching wind speed profile has the shape of the profile in Figure A4-4B (short grass) and is 10 times thicker than the height of the ridge. These profiles are only applicable to ridges up to several hundred feet high and when the temperature effects do not significantly affect the flow. Notice how greatly the profiles change in shape. Obviously, the profiles for flat terrain (Fig. A4-4) are not applicable in hilly country.

How the power in the wind at the typical windmill height changes as the wind blows over a ridge is indicated by the width of the long arrow and the percentage figures inside. In Figure A4-7C (the 50 percent slope ridge), the vertical scale has been expanded for better clarity. The upwind profile is the same as in A and B, but only the lower half of the profile is shown. Notice that in Figures A4-7A and B, the wind goes smoothly over the ridge and, though not shown, essentially the same wind power factors apply on the back side as on the front side. On the ridge with 50 percent slope, the wind does not flow smoothly down the backside. A large region of high turbulence is created as sketched in Figure A4-7C. Note that the increased speed at the crest is greatest for the ridge with the 25 percent slope.

Of course, ridges never have exactly triangular cross-sections. Still, important conclusions can be drawn for windmill placement:

1. Stay as far from the base of the ridge as possible.

2. Use the top of a ridge, if possible, or at least the upper half.
3. If the ridge is steep (greater than 50 percent) be wary when the wind blows from the other side of the ridge. If you believe a high turbulence area exists at times at the site being considered, avoid that area.

Most ridges will not lie at right angles to the prevailing wind. Wind blowing diagonally across the ridge will snake across it in an S-shaped manner and will tend to slow down and speed up less than when blowing at right angles to the ridge. The ridge can be steeper and still not have the high turbulence region on the backside with the diagonal flow.

Effect of Obstacles on Wind Power

Structures and trees are likely to be in the vicinity of a windmill. How close can a windmill be placed to these, and how much penalty is paid if these obstacles cannot be avoided? Repeating and expanding on what was studied in Chapter 3, both loss of wind speed and the wind turbulence downwind of these items should be considered. A rule of thumb that has been used for wind generator placement is: If the location is not well above these obstacles, the windmill should be placed at least 10 heights or tree widths away from the obstacles. This is a reasonable rule, but the results of some available wind measurements are presented here to give a better feeling for how the wind is disrupted when it passes over and around obstacles.

Buildings. Figure A4-8A depicts the essential features of the flow around a simple building. An undisturbed wind speed profile is seen approaching from the left directly toward the long side of the building. On the downwind side of any building, there will be a region where the air cannot flow smoothly, but the wind is extremely gusty and turbulent. In part of this region, the average flow is even in the reverse direction from the main wind. For most houses, sheds, and barns, this region extends 2.5 to 4 building heights (*H* in the figure) from the building. The maximum height of this extreme turbulence zone is about 50 percent greater than the building height. Disturbed wind speed profiles directly in line and downwind of the building are shown. The representative maximum wind speed decrease, turbulence increase, and wind power loss are listed for 5, 10 and 15 building heights downwind. These maximum effects occur at an elevation between 1 and 2 building heights.

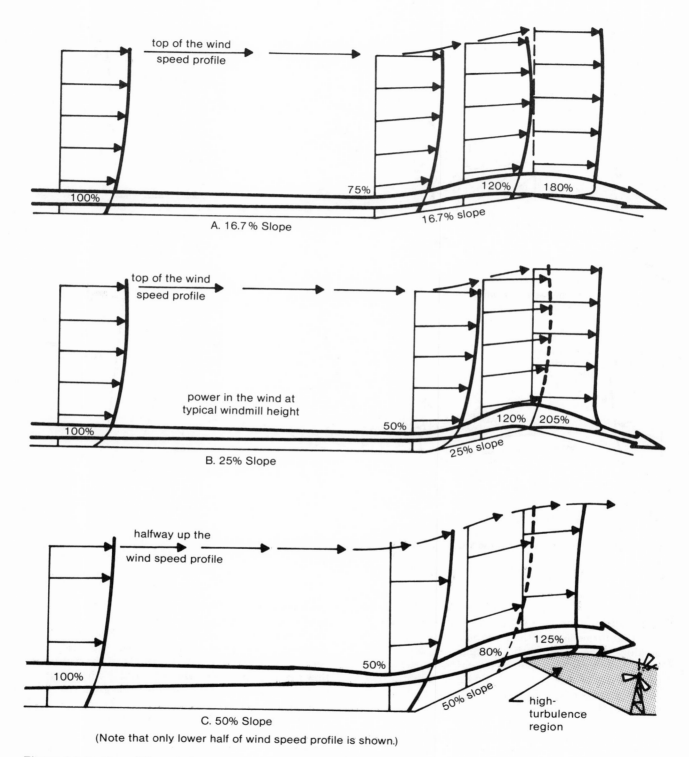

A. 16.7% Slope

B. 25% Slope

C. 50% Slope

(Note that only lower half of wind speed profile is shown.)

Figure A4–7: How different-shaped ridges affect the wind speed profile.

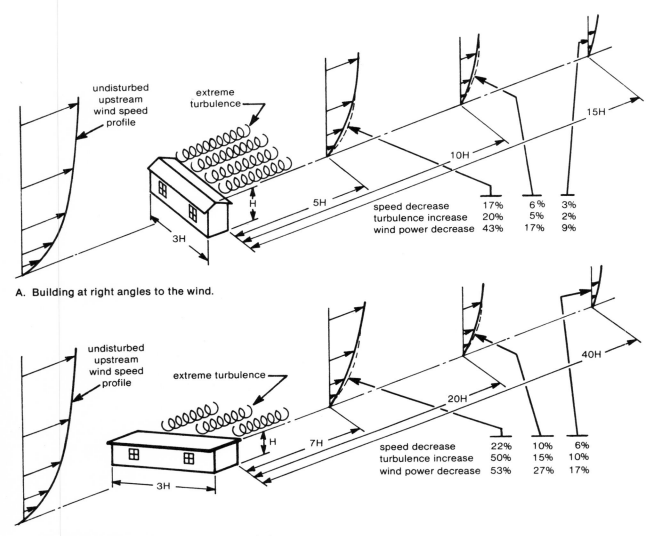

speed decrease	17%	6 %	3%
turbulence increase	20%	5%	2%
wind power decrease	43%	17%	9%

A. Building at right angles to the wind.

speed decrease	22%	10%	6 %
turbulence increase	50%	15%	10%
wind power decrease	53%	27%	17%

B. Flat-topped building at 45 degrees to the wind.

Figure A4–8: How the angle of a building affects wind flow.

Figure A4–8B shows the effect of the wind coming diagonally at a building with a flat roof. A swirling flow that forms around the front and sides of the building is joined by another swirling flow off the front roof corner. The combined swirl is unusually persistent. The effect of this building 40 heights downstream is about the same as 10 heights downstream of the building in A4–8A.

Trees. The disturbed flow region behind the crown of a tree is simpler in nature than that behind a building. At more than about two crown widths downstream the individual disturbances from the branches have blended together. At some distance downwind, the disturbance of a windmill will appear to be much like that of a tree with a crown the same shape as the area sweep by the windmill blades. However, the effect of the windmill will be more persistent. Both the width and height of the disturbed flow region of trees and windmills will gradually grow with distance. The accompanying table gives the growth of the width of the turbulent region and the loss in wind speed and power. The growth rate in height of this turbulent region is about the same as for width. Silos are also included in the table, as they are better treated in this category than as a building. However, notice that the effect of a solid object like a silo is not as persistent as a porous object like a tree.

Extent and Severity of Disturbed Flow Downwind of Trees and Silos

Distance downwind from object[1]	5	10	15	20	30
Width of turbulent flow region (tree widths)		1.5	2	2.5	3.0	3.5
Dense-foliage tree[2] —max. % loss of velocity	20	9	6	4	3
max. % loss of power	49	25	17	13	9
Thin-foliage tree[3] —max. % loss of velocity	16	7	4	3	2
—max. % loss of power	41	18	12	8	6
Silo—max. % loss of velocity	10	5	3	1	—
max. % loss of power	27	14	8	3	—

1. Measured in the number of widths of the object.
2. Colorado spruce-type tree (and windmills, including two-bladed rotors).
3. Pine-type tree.

Rows of trees have often been planted in windy country to shelter crops and conserve soil. Just the presence of these *shelterbelts*, as they are often called, can be an indication of good windmill country. Unfortunately, these tree rows also reduce the wind for windmills. Figure A4-9 shows wind speeds upwind and downwind of two long rows of trees, one with dense foliage and one with loose foliage. The speeds are shown for four heights in terms of percentage of the undisturbed wind speed. Distances are in the number of heights of the trees. Whether the tree row has loose or dense foliage, at 25 heights downwind the wind speed is 90 percent of its undisturbed value at all four heights. The power available there is only 73

percent ($0.90 \times 0.90 \times 0.90 \times 100$) the undisturbed value. Also notice the 5 percent greater wind speed at 2½ heights behind and 1½ heights above the shelterbelt—a good windmill location.

The above results are only for long rows of trees with the wind at a right angle to the row. The wind probably returns to its undisturbed condition in about half the distance if the shelterbelt is only twice as long as it is high. The wind coming at an angle to a dense shelterbelt suffers a considerably reduced disturbance, while the disturbance will be increased for a loose row of trees, at least until an angle of 45° to 60° from the right angle is reached. The reduced wind energy downwind of a shelterbelt shows up as a greatly increased level of turbulence.

A Better Method for Determining Your Long-Term Average Wind Power

A very simple method for estimating your average annual wind power was presented in Chapter 3. A better approach is described here which uses the same simple equipment, a wind cup anemometer with a run-of-the-wind meter. This method involves taking frequent wind run data—probably several times a day—on the hour or nearly so. Using a table, such as the sample in Figure A4-10, the date, hour, and miles of wind measured are entered into columns 1, 2, and 4. Column 3, time lapse, is the difference in time between the latest reading and the previous one. Similarly, column 5, miles of wind difference, is the difference between the latest miles of wind value in column 4 from the previous value.

When you obtain copies of the WBAN 10A sheets (see Chapter 3) from the weather station for days you have kept data, sum the wind speeds for the same time periods and enter in column 6. To do this prop-

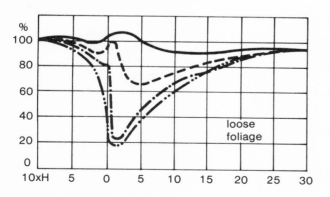

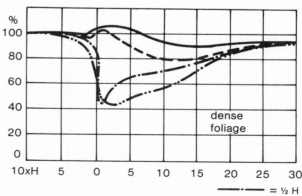

Figure A4-9: Percent wind speed at different heights above surface behind a row of trees.

1	2	3	4	5	6	7	8	9
			YOUR WIND DATA			WEATHER STATION DATA		
		Time lapse	Miles of wind:		Sum of wind speeds during time span:		Predominant wind direction	Wind factor
Day	Time		reading	difference	knots	miles of wind		
						1.15 x ⑥ *		③×⑤÷⑦*
3	8A	—	2170					
	10A	2	2192	22	15	17.3	NW	2.5
	1P	3	2211	19	8	9.2	W	6.2
	4P	3	2238	27	33	38	S	2.1
	7P	3	2280	42	46	52.9	S	2.4
	10P	3	2325	45	51	58.7	S	2.3
4	8A	10	2447	122	—	—	None	—
	11A	3	2486	39	27	31.1	NW	3.8
	1P	2	2494	8	6	6.9	NW	2.3
	5P	4	2506	12	—	—	None	—
		* Circled numbers indicate that you use the value from that column, i.e. 1.15 × 15 = 17.3						

A4-10: Sample comparison of site survey and weather station winds.

erly, add half the wind speed for the beginning and ending hours to all the readings in between. If you have readings only an hour apart, then sum half of each of the two weather station wind readings. For instance, if you have readings two hours apart and the weather station hourly wind speeds for that period are 10, 8, and 11 knots, the value to enter in column 7 is (10 ÷ 2 + 8 + 11 ÷ 2) = 18.5. If your readings are an hour apart and their wind speeds are 8 and 11, (8 ÷ 2 + 11 ÷ 2) = 9.5. To convert to miles per hour, multiply by 1.15 and enter in column 7 (the circled column number in the equation in the heading means "use the value in that column"). The predominant wind direction is entered in column 8. If there is none, note that. A wind factor is determined by multiplying column 3 times column 5, divided by column 7. This value is entered in column 9. Example numbers have been used in Figure A4-10.

All this work is being done to separate by direction the ratio of your wind speed to the wind speed at the weather station. You are making corrections in this process for all the terrain factors that affect your anemometer and the anemometer at the weather station. The next step is to obtain your average wind speed ratio and the weather station wind power ratio for each direction. By combining these numbers in a clever way you can give proper weight to the wind energy from each direction. Set up a table following the headings in Figure A4-11. Under the first column on the left enter the wind directions, divided into eight directions as shown here, or use all 16 directions as given in Figure 3-22 if you have a lot of data and you want to do a better job.

We need to summarize the data collected in the previous table (Figure A4-10) and this information to the new table (Figure A4-11). Separately sum up the column 3 and corresponding column 9 values in the previous table which have the same wind direction (or combined wind directions) and enter these values on the appropriate line in the new table under columns D and E. For each line in the table the wind ratio of column F is obtained by dividing column D by column E (see the examples in Figure A4-11) and the power ratio, column G, is the cube of column F. Stop to look at these numbers. You have completed a comparison of how your wind speed and power compare with the values from the weather station for each direction. Do you need more data samples from several directions to give more confidence in the results? Are obstacles greatly reducing the potential wind power from certain directions?

The clever way referred to earlier of combining the power ratios calculated in column G is to give each value the proper emphasis, based upon the long-term wind averages at the weather station. The key is in filling out column B in the table (Figure A4-11) using the weather station wind summary information sheet as shown in Figure 3-22. Combine the numbers in the last two columns of that form as indicated: (%) times the cube of the *mean wind speed*. If each two wind directions have been combined as in Figure A4-11, add the two resulting values and enter in column B. For example, the first value in column B is obtained from Figure 3-22 as follows:

$$4.8 \times 15.8^3 + 3.8 \times 19.8^3 = 48,400$$

RESULTS FROM WEATHER STATION SUMMARY SHEET (see example, Figure 3-22)			YOUR DATA, EACH DIRECTION (see example, Figure A2-11)				
A	B	C	D	E	F	G	H
Wind Directions	Sum of (%) x (mean wind speed)^3	Directional Wind Power Ratio x Ⓑ *	Sum of Time Spans	Sum of Wind Factors	Wind Ratio Ⓔ÷Ⓓ *	Power Ratio Ⓕ^{3}	Wind Power Ⓒxⓖ
N–NNE	48,400	63	31	27.9	0.90	0.729	46
NE–ENE	240,700	315	26	25.5	0.98	0.941	296
E–ESE	288,900	378	36	40.3	1.12	1.405	531
SE–SSE	249,200	326	56	58.2	1.04	1.125	367
S–SSW	155,900	204	28	24.1	0.86	0.636	130
SW–WSW	160,200	210	35	26.6	0.76	0.439	92
W–WNW	113,900	150	29	26.4	0.91	0.754	114
NW–NNW	90,200	118	37	31.5	0.85	0.614	73
TOTAL	1,347,400						1648

1764 ÷ 1,347,400 = 0.00131 = Ratio

Weather station wind power (from App. 1)

Your long-term wind power, watts per square meter ⟶

* Circled letters indicate that you use the value from that column i.e. E ÷ D = 27.9 ÷ 31 = 0.90

A4–11: Sample table for calculating your long-term average wind power.

Add up all the values in column B. Extract the wind power for the weather station you are working with from Appendix 1 (use the same month or the yearly value to correspond with the data used for column B) and enter on the next line below. Divide this value by the total as indicated in Figure A4–11. The resulting ratio is multiplied into each value in column B to obtain the values for column C. You have proportioned the weather station wind power by direction, thus the heading "Directional Wind Power." The sum of column C equals the average wind power, 1764 watts per square meter in this example.

Finally, fill in column H by multiplying together columns C and G. The total of column H, entered at the bottom of the column is your total wind power in watts per square meter.

If data from another weather station is appropriate for comparison, repeating the whole process of comparing your data to theirs can give you more confidence in your calculated wind power, if the results are close. If the two answers are quite different, possibly a meterologist can help interpret the meaning of the differences. A third weather station comparison may be appropriate.

To summarize, this procedure takes your data from a rather short period of time and compares it with the local weather station data for the same hourly periods (Figure A4–11). Then, this procedure compares the long-term averaged data (for the same weather station) to the average power ratio for each direction to obtain your long-term averaged wind power.

Notice that we have not made use of all the information in the weather station wind summary, Figure 3–22. It has 11 speed ranges with the percent time the wind is in each range and from each direction. This information can be used if a more sophisticated wind speed recorder that tallies the wind speeds by intensity is employed in the site survey. Consult the manufacturer about how to adapt the above procedure to obtain your long-term wind power.

Wind Power Catalog

The following pages show products and companies that are active in the field of wind power. Naturally, this is not a complete catalog—many excellent products are not represented here, and more are coming on the market all the time.

All information in this catalog section was supplied by manufacturers or organizations; we have briefly summarized what is available from each. Our goal is to help you locate the products and information that will allow you to evaluate for yourself the feasibility of a home wind power system.

In using the wind generator comparison chart, keep in mind that prices change constantly. If you need more information about any of the products described in the catalog section, we ask that you write directly to the companies you are interested in.

Also included in this catalog are additional sources of information, descriptions of federal wind energy programs and names and addresses of state energy offices and solar energy associations.

Wind Generators

Aermotor/Division
of Valley Industries, Inc.
Industrial Park
P.O. Box 1364
Conway, Arkansas 72032
(501) 329-9811

Aermotor manufactures windmills for water pumping and
expects to manufacture an electric generator in 1982. Their
multi-bladed mills are available with diameters of 6, 8, 10,
12, 14, and 16 feet. They feature a positive oiling system,
adjustable stroke, and automatic regulation.

Aermotor also makes four-post standard and widespread towers from 21 to 47 feet high and three- or four-
post stub towers from 3 to 20 feet high for use with
existing towers.

Aero Power Systems, Inc.
2398 4th St.
Berkeley, California 94710
(415) 848-2710

Aero Power Systems, Inc. manufactures the *SL1500 Wind
Powered Electric System.* The wind generator is a three-
blade spruce propellor with stainless steel leading edges.
Blades feather automatically at high wind speeds. A tail
vane controls the position of the propeller and prevents
damage from gusts and turbulence.

The drive system is a high-torque drive (HTD)® that
eliminates the gears-in-oil-drive method formerly used by
APSI. The HTD® system reduces noise, cost and maintenance. Aluminum and stainless steel is used throughout,
making the unit ideal for highly corrosive environments.
The unit puts out 1500 watts at 22 mph. Solid state automatic voltage regulations and controls for overcharge protection are part of this system.

The SL1500 comes in five configurations: four battery
chargers, 12, 24, 48 and 110 volts DC; and also as a utility
intertie when used in conjunction with a specially built and
matched, efficient synchronous inverter. The SL1500 has a
12-foot rotor; cut in wind speed is 7.5 mph; prices are
$3300 to $4000.

Ampair Products
Aston House
Blackheath
Guildford
Surrey GU4 8RD
England

Ampair manufactures the *Ampair 50* wind driven alternator
for marine applications. It is designed for permanent installation on cruising yachts.

The wind turbine has 14 individually replaceable polypro-
pylene blades, mounted in an aluminum hub. It can be
used with 14 blades for low wind speed environments or 7
for higher wind speed environments. It is 26 inches in

Ampair 50

diameter and includes a 30-inch-long mounting tube.

The alternator is capable of operation at very low
speeds. It includes a permanent magnet rotor and remotely
located rectifier for converting AC to DC.

The Ampair 50 can produce up to 50 watts of 12 or 24
volt DC electricity. Voltage limiter and read-out unit are optional. A special mounting bracket kit costs extra.

Bergey Wind Generators
c/o Windward, Inc.
8 River Drive
Hadley, Massachusetts 01035
(413) 584-3510

Windward, Inc. of Hadley, Massachusetts markets a complete homesite wind system that includes the *Bergey BWC
1000-S Wind Generator.* The cost of the system, including
the 1000-watt generator, a 60-foot tower, site analysis and
several accessories, is $5,299. The BWC is an upwind
horizontal axis windpower generator. It reaches its rated
output of 1,000 watts at a wind speed of 25 mph. The unit
consists of: a three-blade, all metal rotor system with a
diameter of 8.3 feet; a low-speed, permanent magnet alternator; a main-frame structure; a self-furling tail; rectifiers
and associated wiring; and a synchronous inverter. A
battery-charging system is also available.

Chalk Wind Systems
(formerly American Wind Turbine)
P.O. Box 446
St. Cloud, Florida 32769
(305) 892-7338

Tom Chalk's recently invented, multi-blade windmill is
custom built for specific applications. Its aluminum blades
are attached at the outer perimeter to an aluminum rim.
Aluminum wire spokes like those of a bicycle run from the
hub to the rim to create a strong framework for the mill. Its
airfoil profile allows it to run at winds as low as 3 mph. It is
protected against high winds by a tail vane.

It can be used to pump water (82 gallons per minute in
16 mph wind) or to generate electricity by means of an
alternator run from a belt on the rim. It can provide 700
watts in 20 mph winds.

Coulson Wind Electric, Inc.
RFD 1 Box 225
Polk City, Iowa 50226
(515) 984-6038

Coulson Wind Electric, Inc. sells and services new, used and rebuilt wind equipment, towers and batteries. Coulson's primary product is a *Zephyr Model "B"* wind turbine system. The system is driven with either a four- or an eight-blade wooden propeller. The 21- or 25-foot diameter rotor is designed to produce 15 kilowatts at 25 mph. The output is 120, 208, 240 or 440 volts AC, three phase or single phase. Coulson also sells 8-, 10- and 12-foot laminated wood replacement propellers for older *Wincharger* models.

Coulson Zephyr Model B

Dempster® Industries, Inc.
Beatrice, Nebraska 68310
(402) 223-4026

Dempster® manufactures *No. 12 Annu-Oiled Windmills* for pumping water. They are multi-blade mills in 6, 8, 10, 14, and 18 foot sizes. The blades are made of heavy gauge steel and fastened to circles to keep proper curvature and position. Machine-cut equalizing gears run in an oil bath that needs changing only once a year. Gears and other working parts are housed in a weatherproof hood made of galvanized sheet steel.

Dempster steel towers with anchor and wood pump poles are available at an additional cost. They come in heights of 22, 28, 30, 33, 39, and 40 feet.

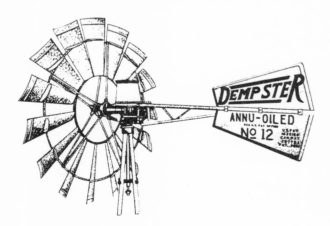

Dunlite
Head Office, Manufacturing, and Export Division
28 Orsmond St.
P.O. Box 100
Hindmarch, South Australia 5007
61 846 3832

Dunlite has manufactured a high-quality, 2-kilowatt wind generator for more than 40 years. The *Dunlite 2000W* has a maximum continuous output of 2000 watts at 25 mph winds. The three-blade, galvanized steel propeller is governed by a feathering mechanism in winds up to 80 mph. The gear-driven, brushless alternator cuts in at 10 mph. The Dunlite 2000W includes voltage regulator and control box.

In the United States, Dunlites are sold by Enertech, Inc., P.O. Box 420, Norwich, Vermont 05055; (802) 649-1145.

Dynergy Corp.
P.O. Box 428
Laconia, New Hampshire 03246
(603) 524-8313

Dynergy manufactures a three-blade, vertical-axis Darieus rotor. Among the options available with this system are: mechanical coupling, chain and sprocket coupling; speed increaser or reducer to power devices such as pumps; synchronous inverter to permit the user to "tie-in" to a utility power grid; and a gear box matched to average annual wind speed for more efficient wind use. Advantages of the Dynergy wind turbines include: it responds to wind from any direction so tail vanes are not required to aim it into the wind; expensive blade-pitch controls are not necessary; the turbines may be "stacked" or installed one on top of another to achieve greater output.

Elektro G.m.b.H.
St. Gallerstrasse 27
CH-8400 Winterthur
Switzerland

Since 1938, Elektro has been manufacturing a variety of wind turbines with interchangeable parts that withstand severe climatic conditions. They all run at low speeds and have a long service life, sometimes 30 to 40 years.

The vertical-axis, six-blade turbine is ideal for areas with high windspeeds. Two types are available: *Model W50* has a maximum output of 50 watts; and *Model 250* has a maximum output of 150 watts. Both can withstand windspeeds of more than 60 mph and do not need an automatic control.

The horizontal-axis, propeller-type turbine has two or three blades and is available with maximum outputs of 600 to 10,000 watts. The 6000- and 10,000-watt models can supply electricity for an average home (excluding cooking and hot water heating, which can be done with gas). They can be used with or without a handbrake.

All Elektros have wooden blades governed by centrifugal weights and adjustable tail fins and use field or permanent magnet alternators. Models up to 6 kilowatts are direct drive.

Elektro also manufactures wind generators for heating only and sells towers, batteries, inverters, and controls separately or in systems with their turbines. A recently published manual, *Thirty-Seven Years Experience with Elektro Windmills*, explains their products and history ($30.50).

Canadian distributor:
Future Resources and Energy Ltd.
P.O. Box 1358
Station B
Downsview, Ontario M3H 5W3
(416) 630-8343

American distributor:
Real Gas and Electric Co.
P.O. Box 193
Shingletown, California 96088
(916) 474-3852

Enertech
P.O. Box 420
Norwich, Vermont 05055
(802) 649-1145

Enertech is a leading manufacturer and supplier of wind-powered generating systems. The company's generator, *Enertech 1800*, supplies 1800 watts at 24 mph. The unit has a 13-foot diameter rotor with three wood blades, a fiberglass nacelle and spinner and rotor overspeed controls. The generator de-energizes and the brake locks on automatically upon loss of utility power. The generator produces 115-volt AC 60-cycle electricity at all operating wind-speeds. Because this electricity is identical to utility power, the Enertech 1800 can be connected to a 30-amp. dedicated circuit and the electricity it generates will flow directly to any lights or household appliances on the house-side of the utility meter. The company also sells a larger 4000-watt system and several models manufactured by Sencenbaugh, Dyna Technology and Dunlite. Small Sparco windmills for pumping water are also available through Enertech.

Finishing an Enertech installation.

TRADEMARK REGISTERED 1935

Jacobs Wind Electric Co., Inc.
2180 W. 1st St., Suite 410
Ft. Myers, Florida 33901
(813) 334-0339

Jacobs, manufacturer of thousands of wind generators used on farms in the 1930s, 40s and 50s, now produces a new model called the Jacobs *8-10 KVA Wind Energy System*. The company says its system features a special factory-assembled and individually power-tested design that is based upon the company's 50 years of experience. The new model has a 23-foot diameter rotor. The Jacobs system includes a *Mastermind* solid-state control unit that regulates the alternator output at all wind speeds to interface through a line synchronous inverter with the 60-cycle highline current in the home. The company also sells towers and other accessories.

Kedco, Inc.
9016 Aviation Blvd.
Inglewood, California 90301
(213) 776-6636

Kedco manufactures a family of wind generators based on Jack Park's plans for a three-blade model. *Models 1200* (12 foot diameter) and *1600* (16 foot diameter) are both rated at 1200 watts maximum for battery-charging applications. *Models 1205* and *1605*, also for battery-charging, are rated at 1900 watts maximum.

Models 1210 and *1610* are supplied with 2000-watt DC generators. *Model 1620* is supplied with a 3000-watt DC generator of permanent magnet design for synchronous inverter operation, as well as wind furnace and other applications.

The fundamental difference between the *1200* and *1600* series is the blade diameter. The increase in diameter from 12 to 16 feet nearly doubles the energy yield.

All models have aluminum blades and these features: automatic blade feathering; ground shut-off and re-set cables; automatic vibration-sensing shut-off; and gear drive.

The *1200/1600 Owner's Manual* costs $7.50. The price is refundable upon purchase of any complete Kedco wind generator.

Lubing-Maschinenfabrik
Ludwig Bening
2847 Barnstort
Postfach 110
West Germany

Lubing manufactures about 50 different sizes of wind-driven water pumps and only one type of windmill generator—a downwind, horizontal-axis machine. The Lubing mill is rated at 400 watts in a 27 mph wind.

The six-blade Lubing propeller has three small, fixed-pitch blades that start the mill at 9 mph. The three larger variable-pitch blades produce the bulk of the power at higher wind speeds. All the blades are made of epoxy resins reinforced with fiber glass. A centrifugal governor, fitted to each of the variable-pitch blades, prevents them from exceeding the maximum speed of 600 rpm.

Output from the alternator is converted from AC to 24 volts DC at the control panel. The electronic controls regulate the charging of the batteries automatically, and when a battery voltage of 28.5 volts is reached the charging current is cut off.

The basic system includes a three-foot stub tower and control panel. The Lubing is also sold complete with aluminum tubular tower in three sizes. The tower is easy to bolt in place and has the added advantage of being hinged at the base. This feature enables the owner to raise the mill by means of the winch provided with the system.

Canadian distributors:
Budgen and Associates
72 Broadview Ave.
Pointe Claire, Quebec
Canada
(512) 695-4073

Mehrkam Energy Development Co.
RD 2 Box 179E
Hamburg, Pennsylvania 19526
(215) 562-8856

Mehrkam Energy Development Company manufactures several medium-to-large-sized wind machines, including the *Mehrkam 2000* and the *Mehrkam 440*. These have six blades and produce 2000 kilowatts and up to 40 kilowatts, respectively. The company considers the 440 an agricultural and residential wind system, designed to produce average amounts of power in low-to-medium wind velocity areas. The 2000, including tower and generator, has a total weight of 90 tons and is designed for industrial uses. Both systems can operate in above-average winds; both have a cut-out speed of 40 mph. Systems larger than the 2000 are available on request. Owners' manuals are available for $8 postpaid. Allow six to eight weeks for delivery.

Mehrkam 440.

Millville Windmills, Inc.
10335 Old 44 Drive
Millville, California 96062

Millville Windmills, Inc. designs and manufactures several small-to-medium-sized wind energy conversion systems, including upwind and downwind models designed for electricity generation and water pumping. The company's *Model 10-3-AIR Downwind* has a 25-foot diameter rotor. The *Model 10-3-IND Upwind* supplies 240 volts of AC power. Wind generator prices are from $10,000 to $12,000. The company also distributes Dempster water-pumping equipment and inverters, battery chargers and wind-recording instruments.

North Wind Power Company
Box 315
Warren, Vermont 05674
(802) 496-2955

North Wind Power Company became active in wind energy conversion systems development in 1974. The company specializes in the design and manufacture of small systems for remote power requirements and residential applications. North Wind can provide site analysis, system design and components, everything necessary for the production of wind power. The company has received two wind-energy development contracts from the U.S. Department of Energy. The first of these culminated in the *HR2* wind electric system that produces 2 kilowatts at 20 mph and the second contract led to the development of North Wind's utility interface 6-kilowatt unit, the *LI6*. The LI6 is to be introduced in early 1982. Northwind's HR2 is marketed in the United States by:

Energy Unlimited, Inc.
Two Aldwyn Center
Villanova, PA 19085
(215) 525-5215

North Wind generator in feathered position.

Pinson Energy Corp.
Box 7
Marstons Mills, Massachusetts 02648
(617) 428-8535

Pinson Energy Corp. manufactures a wind energy system known as the *Cycloturbine*. This is a vertical axis, straight-blade wind turbine with cyclically pitched blades. The Cycloturbine may be adapted for four uses: with the Cycloturbine interface system, the equipment can be tied directly with a utility grid; the Cycloturbine used with either an alternator or generator can provide electricity to run or supplement domestic hot water or resistive baseboard heating; the Cycloturbine battery system can be used at remote sites to power navigational aids or other systems; and the Cycloturbine mechanical drive system can be used to drive an air compressor, water friction heater, irrigation or water pump.

Product Development Institute
1740 Eber Road
Holland, Ohio 43528
(419) 865-3836

Product Development Institute manufactures a *Wind Jennie Model 6500* wind energy conversion system that produces 6,500 watts at approximately 26 mph. The Model 6500 is an induction-type system and is designed for utility interface. The system has an upwind, horizontal design with a 20.5-foot rotor. Other operational characteristics include: cut-in wind speed, 9 mph; survival speed, estimated 100 mph; maximum power, 8,000 watts.

Sencenbaugh Wind Electric
P.O. Box 11174
Palo Alto, California 94306
(415) 964-1593

Sencenbaugh Wind Electric, started in 1972 as an agent for Dunlite, has now developed its own range of products as follows:

Sencenbaugh Wind Generator *Model 1000-14* is rated 1 kilowatt at 22 mph wind. An upwind horizontal-axis machine with a three-blade propeller 12 feet in diameter, it is constructed of machine-carved Sitka spruce with bonded copper leading edge. The propeller speed is 175 rpm at cut-in and 290 rpm at maximum output. Transmission is through a helical gear with a 3:1 ratio and is over-designed to use only 25 percent of rated capacity at full output. Gear and alternator, a three-phase low-speed type, are both sealed in a cast aluminum body. Maximum continuous output at 14 volts DC is 1,000 watts at 22 to 23 mph; peak output is 1,200 watts. The cut-in and charge rates are electronically controlled by a solid state speed sensor and voltage regulator pioneered by Sencenbaugh in 1973. Propeller overspeed control is provided by a foldable

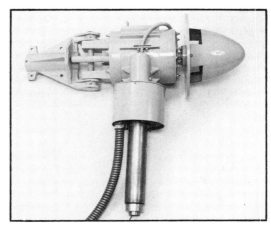
Sencenbaugh generator.

tail at 25 to 30 mph. Maximum design wind speed limit is 80 mph. Package includes control panel.

Sencenbaugh *Model 500-14* is similar to the above, except for the following: The 500-14 has an output of 500 watts at 25 mph, a peak output of 600 watts, and the propeller is only 6 feet in diameter. It is designed for use in severe climates with high average wind speeds. Thus, it has a small propeller, direct drive, and is designed to withstand a maximum wind speed of 120 mph. The new Sencenbaugh *500-14 HDS* is similar to the 500-14, but costs about $1,000 less.

Sencenbaugh *Model 24-14* is a 24-watt wind generator designed to trickle charge 12-volt batteries on boats. The 20-inch diameter propeller direct drives a permanent magnet DC generator. The maximum output is rated at 21 mph.

Sencenbaugh also supplies Rohn towers especially equipped with tower tops for their wind plants in 40, 50, 60, 70, and 80 foot heights.

Sencenbaugh sells inverters, batteries, meteorological instruments and Sitka spruce propeller blades in addition to its own and Dunlite windplants. They will rent their equipment for a deposit of the full list price of the item.

Wind Generators Compared

| Manufacturer | Model | Blades | | Rated Output |
		No.	Diameter	
Aermotor	802	18	6 ft.	
		18	8 ft.	
		18	10 ft.	
		18	12 ft.	
		18	14 ft.	
		18	16 ft.	
Aero Power	SL1500 Wind-powered Electric System	3	12 ft.	1,500 watts @ 22 mph
Ampair Products	Ampair 50	14	26 in.	50 watts
Bergey	BWC 1000-S	3	8.3 ft.	1,000 watts @ 25 mph

Winco/Division of Dyna Technology, Inc.
7850 Metro Parkway
Minneapolis, Minnesota 55420
(612) 853-8400

The Winco *Wincharger®* is a 12-volt generator suitable for minimum lighting in a cabin or campsite. The mill is a two-blade, propeller type, six feet in diameter, with a patented air brake governor that operates by centrifugal force in winds over 23 mph. It direct drives a DC generator, starting at 7 mph. It has a maximum capacity of 200 watts and can generate 20 kilowatt hours (kWh) per month at 10 mph, 23 kWh per month at 12 mph, and 30 kWh per month at 14 mph.

A 10-foot stub tower and control panel are included. Voltage regulator costs about $80 extra.

Winco® also makes three series of electric generators: a basic portable power series; a high-performance series with automatic idler and superior motor starting ability; and a heavy-duty, continuous-power series using slow speeds for long life and full economy.

The Wind Energy Supply Company Ltd.
Iroko House
Bolney Ave.
Peacehaven
Sussex BN9 8HF
England

The Wind Energy Supply Company makes *Wesco® System A Electricity Generating Windmills*. System A windmills feature three-blade rotors operating upwind of the tower; automatic feathering; octagonal-section tower; optional guy wire bracing system; and electrical energy conversion options, including electric-resistance heating, water pumping, and battery charging.

Two models are available. *Model A30* has a rotor diameter of approximately 10 feet and a rated output of 1800 watts in a 22 mph wind; *Model A55* has a rotor diameter of approximately 18 feet and a rated output of 6500 watts in a 22 mph wind. Both cut in at 8 to 11 mph winds and can survive winds up to 125 mph. They are gear driven.

Windworks, Inc.
Route 3, Box 44A
Mukwonago, Wisconsin 53149
(414) 363-4088

Windworks manufactures several wind energy products, including the *Windworker 10*, a system with a rated capacity of 10 kilowatts at a wind speed of 10 mph. This is a three-blade, downwind rotor that achieves full power output at 150 rpm. Output is converted to line voltage and frequency by a Gemini synchronous inverter. The rotor is designed for 200,000 hours of operation (25-year life), and can survive a 165-mph wind. A hydraulic governor and pitch control system regulate blade pitch to accommodate various wind speeds. The cut-in wind speed is 8 mph. Typical annual energy output is 20,000 kilowatt hours at 10 mph average wind speed and 60,000 kilowatt hours at 20 mph. Direct all inquiries to:

Energy Unlimited, Inc.
Two Aldwyn Center
Villanova, PA 19085
(215) 525-5215

Zephyr Wind Dynamo Co.
P.O. Box 241
Brunswick, Maine 04011
(207) 725-6534

Zephyr manufactures a single-blade wind system known as the *OBT 5*. This model, with an output of 150 watts at 25 mph, is especially designed for small applications, camps, remote sites, radio repeaters, recreational vehicles and school demonstrations. The company also manufactures the *Tetrahelix*, a slant-axis turbine designed for battery trickle charging. Other products include a series of alternators, inverters and other accessories (see Alternators and Inverters).

Volts	Cut In	Price	Features
	7 mph	$755–$7,500	For pumping water; 802 parts interchangeable with 702 series parts. Positive oiling system. Adjustable stroke. Automatic regulation. Towers and stub towers available. Prices include windmill only.
12/24/48/110	7.5 mph	$3,300–$4,000	Spruce blades Includes gear-driven alternator, mill, voltage regulator, controls, batteries. Towers available.
12	11–18 mph	$600–$700	For permanent installation on cruising yachts. Suitable for low or average windspeeds. Includes mounting tube, alternator with rectifier, and mill. Mounting equipment, voltage regulator, and read-out unit cost extra.
	10 mph	$5,299	Price includes generator, tower, accessories, and site analysis.

Manufacturer	Model	Blades No.	Blades Diameter	Rated Output
Chalk Wind Systems	12W (water) 12E (electrical)	36	12 ft.	18 gal hr. @ 10 mph 1,000 watts @ 28.30 mph
Coupson Wind Electric, Inc.	Zephyr model "B"	4 21 8	25 ft.	15,000 watts @ 25 mph
Dempster Industries, Inc.	No. 12 Annu-Oiled Windmill		6 ft. 8 ft. 10 ft. 12 ft. 14 ft.	
Dunlite	Dunlite 2000W	3	13 ft.	2,000 watts @ 25 mph
Dynergy	Dynergy	3 vert. axis	15 ft. (equatorial diameter)	15,000 to 35,000 watts
Elektro G.m.b.H.	W 50 W250	6 6		50 watts 150 watts
	WV 05 WV 15G WV 25G WV 35G WVG 50G WVG 120G	2 2 2 2 2 2	9 ft. 10 in. 12 ft. 14 ft. 5 in. 16 ft. 5 in. 22 ft.	600 watts 1,200 watts 4,000 watts 6,000 watts 10,000 watts
Enertech	Enertech 1800	3	13 ft.	
Jacobs	Jacobs 8-10 KVA	3	23 ft.	10,000 watts @ 25 mph
Kedco, Inc.	1200 1600	3 3	12 ft. 16 ft.	1,200 watts @ 21-22 mph 1,200 watts @ 16-17 mph
	1205 1605	3 3	12 ft. 16 ft.	1,900 watts @ 24-25 mph 1,900 watts @ 20-21 mph
	1210 1610	3 3	12 ft. 16 ft.	2,000 watts @ 25-26 mph 2,000 watts @ 21-22 mph
	1620 1848	3 3	16 ft. 18 ft.	3,000 watts @ 25-26 mph 5,000 watts @ 25-27 mph
Lubing Maschinenfabrik		6		400 watts @ 27 mph
Mehrkam Energy Development Co.	Mehrkam 440	6	37 ft.	20 kilowatts @ 25 mph
	Mehrkam 2000	6	160 ft.	2,000 kilowatts @ 25-40 mph
Millville Windmills, Inc.	10-3-IND	3	24.3 ft.	10,000 watts @ 25 mph
North Wind Power Co.	HR2	3	16.4 ft.	2,000 watts @ 20 mph
	LI6	2	33 ft.	6,000 watts @ 20 mph

Volts	Cut In	Price	Features
	3–4 mph 8–9 mph	$1,475–$1,950	For pumping water or generating electricity by belt-drive alternator. Custom built. Aluminum blades attached to wire spokes.
120/208/240/440	8–9 mph	$14,600	Price includes wind generator cost only.
		$ 775 $ 1,150 $ 2,040 $ 3,390 $ 4,860	For pumping water. Steel blades. Positive oiling system
24/32/48/110	10 mph	$3,500	Steel blades. Includes mill, voltage regulator, control box, gear-driven alternator.
115/230	10 mph	$7,250–$12,400	Exact price and specifications depend upon system and options selected.
6/12/24 12/24/36			Vertical-axis. Can withstand high windspeeds. Towers, batteries, inverters, and controls available.
12/24/36/48 12/24/36/48 24/36/48/115 48/60/115 60/115 115			Wooden blades. Includes mill, direct-driven alternator. Towers, batteries, inverters, and controls available.
115	8 mph	$3,950	A successor to the Enertech 1500.
220–240	5 mph	$14,400	System has an automatic storm protection control; price includes inversion system.
12 12	6–8 mph 7–9 mph	$ 3,295 $ 3,895	For battery charging. Aluminum blades.
24 24	6–8 mph 7–9 mph	$3345 $3945	For battery-charging.
Variable 180 maximum	10–12 mph 9–11 mph	$3,595 $4,195	Supplied with dc generator.
Variable 180 maximum 48	10–12 mph 10–12 mph	$4495 $4975	Supplied with dc generator.
24	9 mph		Blades made of epoxy resins reinforced with fiberglass includes controls, mill, alternator, three-foot sub tower. Full-size aluminum tubular towers available.
—	7 mph		Intended for residential, agricultural, and industrial uses.
—	5 mph		
240	9 mph	$10,000–$12,000	Model series may be equipped with induction generator, air compressor or variable-frequency generator.
24, 32, 48, 120	8 mph	$8,700	
240	8 mph		LI6 available for sale in early 1982.

Manufacturer	Model	Blades No.	Diameter	Rated Output
Pinson Energy Corp.	Cycloturbine	3	16 ft.	5,000 watts @ 24 mph
Product Development Institute	Wind Jennie Model 6500	3	20.5 ft.	6,500 watts @ 26 mph
Sencenbaugh Wind Electric	1000-14	3	12 ft.	1,000 watts @ 22–23 mph
	500-14	3	6 ft.	500 watts @ 25 mph
	500-14 HDS	3	6.5 ft.	500 watts @ 24 mph
	24-14	3	20 in.	24 watts @ 21 mph
Winco (Dyna Technology)	Wincharger	2	6 ft.	200 watts @ 23 mph
The Wind Energy Supply Co., Ltd.	Wesco System A windmills: Model A30 Model A55	3 3	10 ft. 18 ft.	1,800 watts @ 22 mph 6,500 watts @ 22 mph
Windworks, Inc.	Windworker 10	3	33 ft.	10,000 watts @ 22 mph
Zephyr	OBT 5	1	5 ft.	150 watts @ 25 mph

Wind Instruments and Towers

Dwyer Instruments, Inc.
P.O. Box 373
Junction Ind. 212 and U.S. 12
Michigan City, Indiana 46360
(219) 872-9141
(312) 733-7883

Dwyer manufactures small, inexpensive wind-measuring devices. The Dwyer *Hand-Held Wind Meter* gives an accurate direct reading in mph for both high (4–66 mph) and low (2–10 mph) wind speeds. It comes in a clear plastic case and costs $8.95.

The *Mark II Windspeed Indicator* consists of a wall-mounted indicator panel, which indicates windspeed in 0

Dwyer Mark II wind speed indicator.

to 80 mph and in 1 to 12 Beaufort Scale, and a rooftop vane that acts as a sensor. The complete package also includes 50 feet of flexible, double-column tubing to be connected to the rooftop vane, mounting hardware, fluid, and instructions. It costs $36.95.

Edmund Scientific Co.
101 East Gloucester Pike
Barrington, New Jersey 08007
(609) 547-3488

Edmund Scientific sells several products related to wind and other forms of alternate energy. These include wind-measuring equipment and the *Wincharger Windmill*, that generates 12 volts of DC power (see Wind Generators).

Natural Power, Inc.
Francestown Turnpike
New Boston, New Hampshire 03070
(603) 487-2426

Natural Power manufactures instruments and components for wind energy systems. Their wind survey and site analysis instruments include windspeed and/or wind direction recording instruments (single-channel, $360–$395, dual-

Volts	Cut In	Price	Features
140/240	8 mph	$7,000–$9,000	Price includes 60-foot, Unarco-Rohn free-standing tower; vertical-axis, straight-bladed wind turbine.
240		$9,282	Three-year conditional warranty.
12/24	6–8 mph	$3,450	Includes mill, gear-driven alternator, controls.
12/24	9–10 mph	$2,850	Designed for severe climates. Includes mill, direct-driven alternator, controls.
14	8–9 mph	$1,600	
12	8–9 mph	$750	For trickle-charging batteries on boats. Deck-mounted. Includes mill, generator, controls.
12	7 mph		For light loads and vacation homes. Includes mill, direct driven generator, controls, 10 ft. stub tower. Voltage regulator costs extra.
	8–11 mph	$4,406–$9,744	For generating electricity. Electrical energy conversion options include electric resistance heating, water pumping, battery charging.
110/240	8 mph	$30,000	Three-blade, extruded aluminum, downwind rotor; price includes 60-foot tower, Gemini inverter.
12	7 mph		Single-blade turbine for small power applications.

channel, $855–$1,035), a wind data accumulator ($180), windspeed compilators ($925–$2,280, depending on number of anemometers), and wind speed compilators with one anemometer and one wind direction head ($1,650). Separate anemometer and wind direction heads for use with these instruments cost $80 each.

Natural Power components for wind energy systems are equipped for various voltages. They include: a generator control panel for a DC generator used to charge a battery pack; a dynamic loading switch used to prevent overcharging and underdraining and to choose appropriate modes of electricity supply (from turbine, storage, or alternate source); a voltage regulator; and a brushless DC alternator.

Generator Control Panel—$495
Dynamic Loading Switch—$140
Voltage Regulator—$85
Alternator—$440–$465

They also stock *Gemini Synchronous Inverters* by Windworks, Inc.

Natural Power instruments and components can either be purchased or leased for a minimum of three months. The company also manufactures solar energy products.

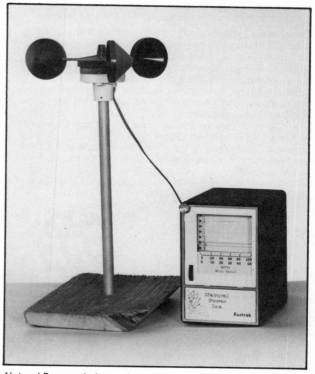

Natural Power wind speed recorder.

Taylor Instruments
Sybron Corporation
Arden, North Carolina 28704
(704) 684-8111

The Taylor *Wind Speed Indicator* is a precision instrument with two scales: 0 to 25 and 0 to 100 in a mahogany case. It is very accurate, especially at the lower ranges. Sold with 60 feet of lead-in wire and instructions, it costs $185.

Unarco-Rohn
Division of Unarco Industries, Inc.
6718 West Plank Road
P.O. Box 2000
Peoria, Illinois 61656
(309) 697-4400

Unarco-Rohn designs and manufactures both guyed and self-supporting towers for use with electric wind generators. Their towers are hot-dip galvanized after fabrication and come with top sections compatible with many wind generators, including Sencenbaugh, Dunlite, Elektro, Jacobs, and Zephyr. Special towers for mounting heights greater than 100 feet and for unusually high wind conditions are also available.

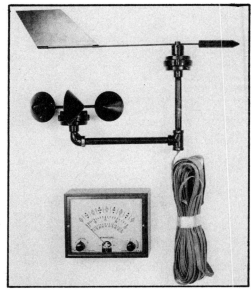

Taylor wind speed indicator.

Alternators and Inverters

Delatron Systems
553 Lively Boulevard
Elk Grove Village, Illinois 60007
(312) 593-2270

Delatron manufactures DC-to-AC inverters. Their VC-type inverters are especially suitable for alternate energy applications, such as wind electric systems. They feature low-idle (no-load) power consumption and high in-rush current-handling capability.

Their VS-type inverters are suitable for applications where a pure sine wave is required, such as for television and telephone operation.

All Delatron inverters feature low DC voltage shut-down and overload shut-down. Their output voltage is 115 to 230 volts AC, 60 hz. They range in size from 3 KVA to 6 KVA. It is advised that they be used wih a DC input voltage of 48 or 120 volts DC.

Delatron also makes deep-cycle, lead acid batteries that can store from 1400 to over 2000 watts for 20 hours.

Georator™ Corporation
P.O. Box 70
9016 Prince Williams St.
Manassas, Virginia 22110
(703) 368-2101

Georator™ manufactures the *Nobrush®* permanent magnet alternator. It is available in frequencies up to 1000 hz and in power ratings from 150 VA to 25 KVA. The rugged Series 36 alternator has stable output voltage, short circuit immunity, high overload capacity, low RF noise, and low heating. It has a maximum operating speed of 600 rpm, with many models operating within 1800 rpm. It can be mounted with shaft horizontal, vertical or at any angle.

Georator™ generator-alternator.

The *Nobrush®* permanent magnet alternator is built to customer specifications with regard to the power, voltage, and frequency at a requested speed.

Natural Power, Inc.

See Wind Instruments and Towers.

West Wind Electronics, Inc.
P.O. Box 542
Durango, Colorado 81301
(303) 884-9709

West Wind manufactures prototype inverters that have been operating since August 1975 with no failures. The 6-kilowatt inverters provide a square wave output of 90 to 140 volts AC. Despite the square wave voltage, no overheating problems have occurred with induction motors or transformers. It is direct coupled (no output transformer); therefore, each AC appliance must have its own fuse in the event of inverter malfunction.

Options include overvoltage and undervoltage shutdown protection; an automatic load demand switch that saves 100 watts continuous and a sine wave filter for TVs, stereos, and other appliances that may not work properly on square wave voltage.

The Voltage Control Switch protects storage batteries from over or undercharging by sensing battery voltage levels and activating power switches used to control various loads.

Two models are available: *Model A* switches on large loads when batteries are charged and energy is still available from the windmill, thus preventing overspeeding; *Model B* starts a back-up generator when batteries are discharged and stops it when batteries are fully or partially charged. (The backup generator must have its own starter motor and logic circuits.) Model B also will switch off the field current of the mill when excess energy is available, thus preventing overcharge.

Windworks, Inc.
Rt. 3
Box 329
Mukwonago, Wisconsin 53149
(414) 363-4088

Windwork's *Gemini Synchronous Inverter* converts voltage from a DC power source, such as a wind turbine, to AC at standard line voltage and frequency. With the Gemini, excess AC voltage can be stored in the public utility grid. When less power is available than is required, the power company can supply it at standard rates, giving credit for excess power fed to it previously. The overall cost of the Gemini Synchronous Inverter is lower per kilowatt than battery/inverter systems, but satisfactory arrangements must be made with the local power company, and the homeowner must be prepared to sacrifice some of his freedom.

The Gemini Synchronous Inverter is available in single-phase systems rated to several capacities. The following gives an example of the costs:

4 kw maximum conversion power	$2,465
8 kw maximum conversion power	$2,855
10 kw maximum conversion power	$4,100
15 kw maximum conversion power	$5,865

Zephyr Wind Dynamo Co.
21 Stanwood Street
Brunswick, Maine 04011
(207) 725-6534

Zephyr Wind Dynamo Company manufactures *Very Low Speed Permanent Magnet Alternators (VLS-PM)*. These direct-drive alternators are suitable for horizontal- and vertical-axis turbines. Their chief advantage is that they do not require shaft-speed increasers. They are built in a modular system that allows costly stator sections to be added or removed while the alternators are in service, providing versatility and easy maintenance. They achieve load matching without power-consuming or switching components. Almost any turbine power curve can be matched. Other features of the *VLS-PM* include heavy-duty bearings, weathertight construction, and solid-state circuitry to provide over-voltage protection. The *Series 331B VLS-PM* has a rated output of 1500 watts at 450 rpm, 180 hz, two-phase. Its voltage is rated at 24 volts AC parallel-connected and 48 volts AC series-connected. Its diameter is 13.9 inches overall and it will support most turbines five meters in diameter and under.

The *Series 647 VLS-PM* is rated at 15 kilowatts at 300 rpm, 270 hz. Buyer specifies voltage and one, two, three, or six phases. It is 27.5 inches in diameter.

Zephyr alternators.

Agents

Alaska Wind & Water Power
P.O. Box 6
Chugiak, Alaska 99567
(907) 688-2896

Active agents for a complete range of wind generators and water turbines.

Alternate Energy Systems
150 Sandwich Street
Plymouth, Massachusetts 02360
(617) 747-0771

Automatic Power
P.O. Box 18738
Houston, Texas 77023
(713) 228-5208

The Big Outdoors People
2201 N.E. Kennedy Street
Minneapolis, Minnesota 55413
(612) 331-5430

Main activity is in geodesic dome housing, but also involved in alternative energy and design of wind power systems.

**Boston Wind
2 Maston Court
Charlestown, Massachusetts 02129**

Boston Wind sells new and used wind generators and accessories and offers an installation service. It also holds courses, workshops and slide-lectures and publishes a quarterly newsletter.

**Budgen & Associates
72 Broadview Avenue
Pointe Claire, P.Q. H9R 3Z4
Canada**

Dr. Harry Budgen is technical advisor on alternative energy to Brace Research Institute of Macdonald College, which has a Lubing, a 5-kilowatt Elektro and a 25-foot diameter sail windmill on campus. Budgen & Associates supply Lubing and Dunlite wind generators and pumps and also plans for a Brace-designed wind generator with three-blade, 32-foot diameter propeller, developing 49.5 hp in 30 mph winds.

**Clean Energy Systems
RD 1, Box 366
Elysburg, Pennsylvania 17824
(717) 799-0008**

Offers mechanical engineering design assistance on wind power and other renewable energy sources.

**Earthmind
5246 Boyer Road
Mariposa, California 95338**

This group of wind power enthusiasts includes Michael Hackleman, author of two excellent books: *Wind & Windspinners* and *The Homebuilt, Wind-Generated Electricity Handbook*. Earthmind deals mainly with reconditioned Jacobs and Winchargers, and since it is a non-profit organization, proceeds from sales go toward establishing its research center.

**Electro Sales Co., Inc.
100 Fellsway West
Somerville, Massachusetts 02145**

Electro Sales supplies surplus inverters, motors, and controls. They are also the distributor for Carter Electric Company, which manufactures rotary converters and inverters.

**Energy Alternatives, Inc.
69 Amherst Road
Leverett, Massachusetts 01054
(413) 549-3644**

These agents for Dunlike, Jacobs, Elektro and Aero Power windmills also offer integrated sun, wind and wood designs.

**Energy Unlimited, Inc.
Two Aldwyn Center
Villanova, Pennsylvania 19085
(215) 525-5215**

Energy Unlimited is the exclusive marketing group for Windworks, Inc. and is the only distributor/marketer in the United States for North Wind's wind-energy system, the HR2. For information on these companies and their products, see Wind Generators.

ENERTECH

**Enertech
P.O. Box 420
Norwich, Vermont 05055
(802) 649-1145**

Enertech manufactures and distributes wind energy systems (see Wind Generators).

Enertech is also the distributor for *Sparco* wind-powered water pumps. These inexpensive, propeller-type wind plants have self-feathering blades. They are easily installed and can be dismounted and drained in freezing weather. In winds of 7 mph, they will pump over 30 gallons of water per hour.

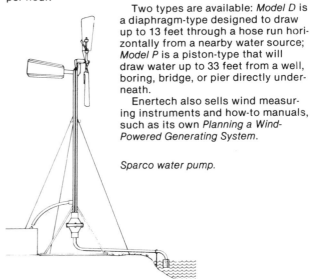

Two types are available: *Model D* is a diaphragm-type designed to draw up to 13 feet through a hose run horizontally from a nearby water source; *Model P* is a piston-type that will draw water up to 33 feet from a well, boring, bridge, or pier directly underneath.

Enertech also sells wind measuring instruments and how-to manuals, such as its own *Planning a Wind-Powered Generating System*.

Sparco water pump.

**Environmental Energies, Inc.
P.O. Box 73
Front Street
Copemish, Michigan 49625
(616) 378-2000**

Founded by Al O'Shea, one of the organizers of the American Wind Energy Association, EEI deals in Dunlite, Winco, rebuilt Jacobs and Elektro machines. It offers a complete installation service and stocks a line of Creative Electronics inverters. EEI practices what it preaches: The shop is powered by an Elektro and is heated by a wood-burning stove and solar energy. Send $5 for their detailed booklet on wind, solar and other energy systems.

**Future Resources & Energy Ltd.
167 Denison Street
Markham, Ontario L3R IB5
Canada
(416) 495-0720**

Fred Drucker of FRE, during his frequent visits to Elektro in Switzerland over the past three years, has developed a thorough understanding of Elektro wind generators. As a result, FRE is now the sole Canadian agent for Elektro. Each machine is checked for quality control before erection. FRE specializes in wind and solar combinations.

Independent Energy Systems
6043 Sterrettania Road
Fair View, Pennsylvania 16415

IES offers reconditioned Jacobs wind generators. Send $2 for information on wind, solar and wood services.

Independent Power Developers
Box 1467
Noxon, Montana 59853
(406) 847-2315

IPD, agent for Dunlite machines, was awarded a contract by the Montana Department of Natural Resources to build and demonstrate a 15-foot diameter, three-blade downwind machine to develop 18 kilowatts in a 33 mph wind. The blades are aluminum.

Jopp Electrical Works
Princeton, Minnesota 55371

Martin Jopp, referred to as a "wind wizard," started back in 1917, when he built a few hundred Jopp Wind generators. His own home has been powered by two 3 kilowatt Jacobs wind generators for years. Jopp now turns out new parts for old Jacobs mills at reasonable prices. He writes about wind systems in *Alternative Sources of Energy*.

Prairie Sun & Wind Co.
4408 62nd Street
Lubbock, Texas 79409

Agent for Winco & Aero Power.

Quirks
33 Fairweather Street
Bellevue Hill
NSW 2023, Australia

Markets the Dunlite wind generator.

Real Gas & Electricity Co.
P.O. Box 193
Shingletown, California 96088
(916) 474-3456
(707) 526-3400 (Santa Rosa office)

Designs and installs systems using Dunlite and Elektro wind generators. Offers complete installation service or supervisory assistance. They also deal with solar and water power.

Rede Corporation
P.O. Box 212
Providence, Rhode Island 02901
(401) 861-5390

Rede is the United States agent for DAF Darrieus wind generators.

Sigma Engineering
Box 5285
Lubbock, Texas 79417
(806) 762-5690

Regional distributor for DAF Darrieus rotors. See Rede Corporation above.

Sunstructures, Inc.
Integrated Architectural Design
201 E. Liberty Street, No. 6
Ann Arbor, Michigan 48104
(313) 994-5650

Specializes in integrated wind and solar systems. Also conducts workshops.

Total Environmental Action
Church Hill
Harrisville,
New Hampshire 03450
(603) 827-3374

Offers a wide range of architectural and engineering services in the alternate energy field. Emphasis on research, design and teaching.

Wind Energy System (Sunflower Power Co.)
Route 1, Box 93-A
Oskaloosa, Kansas 66066
(913) 597-5603

WES sells second-hand wind generators such as Jacobs, Wincharger and Wind King. WES is agent for the Gemini inverter, and designs and builds integrated energy systems. Manager Steve Blake, who has also had extensive experience in building and testing Savonius rotors at the Brace Research Institute, also works with the Appropriate Technology Group at the same address.

Windependence Electric
P.O. Box M 1188
Ann Arbor, Michigan 48106
(313) 769-8469

Windependence sells a range of reconditioned wind generators—Allied, Winpower, Jacobs, etc.

Windward, Inc.
8 River Drive
Hadley, Massachusetts 01035
(413) 323-4940

Windward, Inc. is the New England distributor for a wind generator manufactured by the Bergey Windpower Company of Norman, Oklahoma. This is a small generator that typically produces 2,500 kilowatt hours per year in average winds of 13 mph. Windward supplies a complete equipment and accessory package for the system, as well as other services (see Wind Generators).

Additional Sources of Information

Alternate Sources of Energy
Route 2
Milaca, Minnesota 56353

This very fine journal with many excellent articles on wind power includes "Martin Answers" by Martin Jopp. ASE is *the* journal for the home builder of windmills. It contains lots of solid, safe and intelligent advice. They have published many plans (and by far the best) for home windmill builders. ASE No. 24 (Feb '77) is a special wind power issue. Six issues yearly are well worth the $10 cost.

American Wind Energy Association
1609 Connecticut Ave., N.W.
Washington, D.C. 20009
(202) 667-9137

The AWEA is a national association that represents manufacturers, distributors, and researchers involved in the development of wind energy. Membership is $25 per year. The AWEA issues a quarterly newsletter, publishes the *Wind Technology Journal* and holds convivial and enlightening conferences.

Brace Research Institute
McGill University
Ste. Anne University
Quebec, Canada HOA 1CO

Brace Research Institute publishes booklets and plans concerning alternate energy systems. Their wind energy titles include:

How to Construct a Cheap Wind Machine for Pumping Water, by A. Bodek. $1.25.
Performance Test of a Savonius Rotor, by M.H. Simonds and A. Bodek. $2.
Windmill Power Pumps with Intermediate Electrical Power Transmission, by M.A. Memarzadeh and T.H. Barton. $1.50.
A Simple Electric Transmission System for a Free Running Windmill, by T.H. Barton and K. Repole. $2.
Notes on the Development of the Brace Airscrew Windmill as a Prime Mover, by R.E. Chilcott. 50¢.
Windpower Packet, $1.50. Details of commercially available windmills and short windpower bibliography.

Enclosed payment with order. Add 25¢ if ordering only one leaflet or if paying by check. Make checks payable to the Publication Department.

Earthmind
5246 Boyer Road
Mariposa, California 95338

Wind and Windspinners by Michael Hackleman covers the electronics of wind systems and tells how to build a Savonius rotor.

Farallones Institute
15290 Coleman Valley Road
Occidental, California 94565

Homemade Windmills of Nebraska by Erwin Barbour, originally published in 1898, describes how to build "weird and wonderful" windmills for pumping and sawing.

Flanagan's Plans, Inc.
Box 891
Cathedral Station
New York, New York 10025
(212) 222-4774

Flanagan's sells plans for the patented *Quixote Sailwing Windmill Rotor* designed at Princeton University by staff members of the Flight Concepts Laboratory. The three-blade, horizontal-axis machine has a rating of 1 kilowatt in a 20 mph wind. Maximum wind speed is 50 mph.

The plans include drawings, instructions, and patterns for the rotor, shaft, sails, and bearings. The Quixote is easily constructed by the amateur with locally available materials for about $550. Plans cost $25 and include license for construction.

Forrestal Campus Library
Princeton University
Princeton, New Jersey 08540

The Princeton Sailwing Program by Dr. T. Sweeney is a short report on the two-bladed Sailwing developed at Princeton University. $2.

Four Winds Press
50 West 44th Street
New York, New York 10036

Catch the Wind by Landt and Lisl Dennis is a well-written general introduction to wind power. $7.95.

Great Plains Windustries
Box 126
Lawrence, Kansas 66044

Windustries is a fine regional quarterly newsletter, mainly concerned with wind power. A subscription costs $10; $15 for institutions.

HELION Inc.

Helion, Inc.
Box 445
Brownsville, California 95919
(916) 675-2478

Helion was founded by Jack Park, the author of several books on wind energy and the president of the American Wind Energy Association. Helion publishes *12/16 Construction Plans* for the home construction of a three-blade down-wind turbine. The 12-foot diameter model has a maximum output of 2 kilowatts at 25 mph. A finished model is manufactured by Kedco, Inc. of Inglewood, California. Helion also manufactures microprocessor-based data acquisition systems for energy, environmental and agricultural projects. Helion also provides consulting services and conducts lectures and workshops.

Home Energy Digest
8009—34th Ave. South
Minneapolis, Minnesota 55420

The Home Energy Digest and Wood Burning Quarterly offers practical advice and detailed information to the home-owner interested in alternative energy as a means to self-sufficiency. Articles by experts cover not only wood burning but solar energy, wind power, water power, and energy conservation techniques.

Four issues per year cost $7.95.

Intermediate Technology Publications
9 King Street
London WC2
England

Food from Windmills by Peter Frankael describes the wind-mill-building activities of the American Presbyterian Mission in Ethiopia. Contains lots of good details on how to build sail mills—mainly 11-foot diameter and used for water pumping.

Low Energy Systems
3 Larkfield Gardens
Dublin 6
Ireland
01-960653

Low Energy Systems publishes booklets and offers advice for those interested in alternative energy. In the field of wind energy, they offer the following:

Vertical Axis Sail Windmill Plans. $4. Plans for rotor and tower, suitable for water pumping and grain grinding.
Trickle Charger Windgenerator Plans. $1.50. Suitable for operating light electrical appliances.

National Center for Alternative Technology
Machynlleth
Powys, Wales
England

The *Do-It-Yourself Sail Windmill (Cretan) Plans* show how to build a 12-foot diameter sail wind generator, 200 watt output at 15 mph, maximum 300 watts.

The Bookstore at Natural Power, Inc.
New Boston, New Hampshire 03070
(603) 487-5512

Natural Power sells energy-related publications from their bookstore or by mail. Wind energy titles include:

The Generation of Electricity by Wind Power, by E.W. Golding. $19. A classic textbook including drawings, diagrams, photographs, and bibliography.
Helion Model 12/16 Windmill Plans. $10. (See Helion.)
Electric Power from the Wind, by Henry Clews. $2. A practical guide to individual applications.
Power from the Wind, by P.C. Putnam. $9.95. (See Helion.)
Wind and Windspinners, by Michael Hackleman. $7.50. Step-by-step introduction for the beginner.
Simplified Wind Power Systems for Experimenters, by Jack Park. $6. (See Helion.)
Wind Machines, by Frank Eldridge. $4.25. An overview of the viability, history, and future of wind machines.
The Homebuilt Wind-Generated Electricity Handbook, by Michael Hackleman. $7.50. Covers discovery, restoration, and installation of wind electric machines manufactured in the United States in the 1930's through the 1950's. *Windworks Poster*. $3.50. (See Windworks.)

When ordering, add 75¢ postage and handling. Make check payable to The Bookstore at Natural Power, Inc.

New Alchemy Institute
P.O. Box 432
Woods Hole, Massachusetts 02543

No. 2 of the *Journal of the New Alchemists* contains details of how to build a sailwing rotor. $6.

Publications
Eng. Research Center
Foothills Campus
Colorado State University
Fort Collins, Colorado 80523

Energy from the Wind by Burke and Meroney is the bibliography of bibliographies on wind energy—complete and annotated. $7.50. A "First Supplement" (from 1975 to 1977) is also available for $10.

Rain
2270 N.W. Irving
Portland, Oregon 97210

Rain is a fine monthly "Journal of Appropriate Technology." The April '77 issue contained an excellent article on the 200 kilowatt Dutch Gedser mill—a mill much praised for its low cost and suitability for local manufacture. A *Rain* subscription is $10, a single issue $1.

Structural Clay Products Ltd.
230 High Street
Potters Bar, Herts
England

Reinforced Brickwork Windmill Tower by A.B. Bird concentrates mainly on designing and building brickwork towers. It also contains a section on how to build a 40-foot-diameter sail windmill.

Volunteers in Technical Assistance
3706 Rhode Island Ave.
Mt. Rainier, Maryland 20822

VITA is a nonprofit development organization based in the United States. It supplies information and assistance by mail to people seeking help with technical problems.

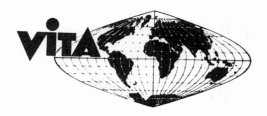

Publications concerning wind energy include the following:

Low Cost Windmill for Developing Nations, by Helmut Bossel. VITA, 1970. $2.95. Instructions for building a windmill with five major components. Can use spare auto parts.
Tanzanian Windmill, by Dick Stanley, VIA/VITA, 1977. $3. Instructions for building a water-pumping windmill developed in Tanzania.
Savonius Rotor Construction: Two Vertical-Axis Wind Machines from Oil Drums, by Jozef A. Kozlowski. VITA, 1977. $3.25. Instructions for building two-stage rotor for pumping water and three-stage rotor for charging auto batteries.

Include payment with order. Make checks payable to VITA Publications Service.
VITA also publishes two newsletters, *VITA News* and *Vis-a-Vis*, containing information on VITA programs, notices of important events, technical abstracts, profiles of VITA volunteers, and lists of problems requiring solutions. Available upon request from VITA Communications Department. Donations accepted.

Wind Energy Report®
P.O. Box 14
Rockville Centre, New York 11571

Wind Energy Report® is an international newsletter devoted to reporting events and trends in the field of wind energy. It features articles by experts, reports on new products, reviews of publications, and news of prominent personalities in the field.
The sixteen-page newsletter is published monthly. A subscription for one year costs $75.

The Federal Wind Energy Program

The main objectives of the federal Wind Energy Program are to accelerate the development of reliable and economical wind energy systems and to encourage the earliest possible commercialization of wind power. The program seeks to advance wind energy technology, to develop a sound industrial technology base, and to address nontechnological barriers to the use of wind energy.

The following laboratories have responsibility for various components of the wind program and publish reports on their various research and development projects. Contact the National Technical Information Service for information on available reports.

These laboratories also issue requests for proposals (RFPs) from time to time and may fund unsolicited proposals which help to achieve program objectives. Each laboratory maintains a mailing list and should be contacted individually.

This list of federal and state energy information sources was compiled by the Solar Energy Research Institute, Golden, Colorado.

**National Aeronautics and
  Space Administration
(NASA) Lewis Research Center**
Provides technical management of large and intermediate system development programs and related supporting research and technology development.

Large Wind Turbine Program
NASA Lewis Research Center
21000 Brook Park Road
Cleveland, Ohio 44135

Publications—Jerry Kennard
(216) 433-4000

Technical information—Ron Thomas
(216) 433-4000

Pacific Northwest Laboratory (PNL)
Provides technical management for wind characteristics research and site validation.

Wind Energy Program
Pacific Northwest Laboratory
Battelle Blvd., P.O. Box 999
Richland, Washington 99352

Publications—Pamela Partch
(509) 942-4410

Technical information—Larry Wendell
(509) 942-4626

Rocky Flats Plant
Manages a test center for commercially available wind systems and administers small machine development and field evaluation programs.

Wind Energy Program
Rockwell International Energy Systems Group
P.O. Box 464
Golden, Colorado 80401

Publications—Darrell Dodge
(303) 441-1300

Technical information—Terry Healy
(303) 441-1300

Sandia Laboratories
Provide technical management of the Vertical-Axis Wind Turbine (VAWT) Machine Development Program and related supporting research and technology development.

Wind Systems Program
Sandia Laboratories
Division 4715
Albuquerque, New Mexico 87185

Publications—Sandia Document
Distribution
(505) 264-3850

Technical information—Emil Kadlec
(505) 264-8669

Solar Energy Research Institute (SERI)

Provides technical management of economic and application studies, investigates institutional issues, administers Wind Energy Innovative Systems Program, and manages technical information dissemination.

Solar Energy Research Institute
1617 Cole Blvd.
Golden, Colorado 80401

Publications—Document Distribution
Service
(303) 231-1158

Technical information—Irwin Vas
(303) 231-1935

U.S. Department of Agriculture (USDA)

Develops requirements for agricultural wind applications and administers related applications research and testing.

Development of Rural and Remote Applications
of Wind Generated Energy
Agricultural Research Services
U.S. Department of Agriculture
Beltsville, Maryland 20705

Publications—
Government & Public Affairs
Information Desk, USDA
Room 100W
14th & Independence
Washington, District of Columbia 20250

Technical information—Louis Liljedahl
(202) 447-3504

National Technical Information Service
U.S. Department of Commerce
5285 Port Royal Road
Springfield, Virginia 22161
(703) 557-4650

The best source of information generated by federal funds. The U.S. Department of Commerce's National Technical Information Service (NTIS) is the central point in the United States for the public sale of government-funded research and development reports and other analyses prepared by federal agencies, their contractors, or grantees. (Ask for the general catalog.)

Requests for publications should be directed to the NTIS operations center in Springfield, Virginia.

An information and sales center for NTIS services and products is located at:

425 Thirteenth St. NW, Room 620
Washington, District of Columbia 20004
(202) 724-3509

State Energy Offices and Solar Energy Associations

Each state government has an energy office. In addition, many states have solar energy associations that are usually nonprofit organizations. These energy offices and associations are good sources of general solar energy information and should be able to make referrals to other sources if they cannot answer specific questions on wind.

Alabama

Alabama Energy Management Board
3734 Atlanta Highway
Montgomery, Alabama 36130
(205) 832-5010

Alabama Solar Energy Association
Johnson Environmental &
Energy Studies Center
University of Alabama at Huntsville
P.O. Box 1247
Huntsville, Alabama 35807
(205) 895-6257

Alaska

Alaska Division of Energy and
Power Development
Mackay Building, 7th Floor
338 Denali St.
Anchorage, Alaska 99501
(907) 276-0508

Alaska Renewable Resources Library
1069 W. 6th Ave.
Anchorage, Alaska 99501

Federation for Community Self-Reliance
P.O. Box 73488
Fairbanks, Alaska 99707

Arizona

Arizona Solar Energy Commission
1700 W. Washington, Room 502
Phoenix, Arizona 85007
(602) 271-3682

Arizona Solar Energy Association
c/o FCCAT, P.O. Box 1443
Flagstaff, Arizona 86002

Arkansas

Arkansas Department of Energy
3000 Kavanaugh Blvd.
Little Rock, Arkansas 72205
(501) 371-1370

California

California Energy Commission
1111 Howe Ave.
Sacramento, California 95825
Public information (916) 920-6430
Publications (916) 920-6216;
in California, toll-free (800) 852-7516

Northern California Solar Energy
Association
P.O. Box 1056
Mountain View, California 94042

Southern California Solar Energy
Association
City Administration Building 11-B,
202 C Street
San Diego, California 92101
(714) 236-0432

Colorado

Colorado Office of Energy Conservation
1600 Downing St.
Denver, Colorado 80203
(303) 839-2507

Colorado Solar Energy Association
P.O. Box 5272
Denver, Colorado 80217

Solar Bookstore
2239 E. Colfax Ave.
Denver, Colorado 80206
(303) 321-1645

Connecticut

Connecticut Office of Policy and
 Management
Energy Division
80 Washington St.,
Hartford, Connecticut 06115
Energy information (203) 566-2800;
 solar information (203) 566-3394

Solar Energy Association
 of Connecticut
Box 541
Hartford, Connecticut 06101
(203) 233-5684

Delaware

Delaware Energy Office
114 W. Water St.
P.O. Box 1401
Dover, Delaware 19901
(302) 678-5644

Florida

Florida State Energy Office
301 Bryant Building
Tallahassee, Florida 32304
(904) 488-6764

Georgia

Georgia Office of Energy Resources
270 Washington St. SW, Suite 615
Atlanta, Georgia 30334
(404) 656-5176

Georgia Solar Energy Association
Campus Box 32748,
Georgia Institute of Technology
Atlanta, Georgia 30332

Hawaii

Hawaii State Energy Office
Department of Planning and
 Economic Development
1164 Bishop St.
Honolulu, Hawaii 96813
(808) 548-4150

Hawaii Natural Energy Institute
2540 Dole Street, Homes Hall
Honolulu, Hawaii 96822

Idaho

Solar Energy Division
Idaho Office of Energy
State Capitol Bldg.
Boise, Idaho 83720
(208) 334-3800

Illinois

Illinois Institute of Natural Resources
Division of Solar Energy and
 Conservation
325 W. Adams St.
Springfield, Illinois 62706
(217) 785-2800

Illinois Solar Energy Association, Inc.
P.O. Box 1592
Aurora, Illinois 60507
(312) 377-9363

South Central Illinois Solar Energy
 Association
c/o Earl G. Powell,
637 Eccles St.
Hillsboro, Illinois 62049
(217) 532-3233

Indiana

Indiana Energy Office
Consolidated Building, 7th Floor
115 N. Pennsylvania Ave.
Indianapolis, Indiana 46204
(317) 633-6753

Hoosier Solar Energy Association
P.O. Box 44448
Indianapolis, Indiana 46202

Iowa

Iowa Energy Policy Council
215 E. 7th St.
Des Moines, Iowa 50319
Solar information (515) 281-8071;
Energy conservation information
 (515) 281-4308

Kansas

Kansas Energy Office
503 Kansas Ave., Room 241
Topeka, Kansas 66603
(913) 296-2496

Kansas Solar Energy Association
c/o Donald R. Stewart
1202 S. Washington
Wichita, Kansas 67211
(316) 262-7427

Kentucky

Kentucky Department of Energy
P.O. Box 11888, Iron Works Pike
Lexington, Kentucky 40511
(606) 252-5535

Kentuckiana Solar Energy Association
c/o David Ross Stevens
Box 974
Louisville, Kentucky 40201
(812) 945-4496

Louisiana

Louisiana Department of
 Natural Resources
Division of Research and Development
P.O. Box 44156
Baton Rouge, Louisiana 70804
(504) 342-4592

Maine

Maine Office of Energy Resources
55 Capitol St.
Augusta, Maine 04330
(207) 289-2196

Maine Solar Energy Association
24 Goff St.
Auburn, Maine 04210
(207) 783-6466

Maryland

Maryland Energy Office
301 W. Preston St., Suite 1302
Baltimore, Maryland 21201
(301) 383-6810

Massachusetts

Massachusetts Office of
 Energy Resources
Solar information
73 Tremont St., Room 700
Boston, Massachusetts 02129
(617) 727-4732

Massachusetts Bay Chapter
New England Solar Energy Association
55 Chester St.
Newton, Massachusetts 02161
(617) 547-1942

Western Massachusetts Solar Energy
 Association
c/o Cooperative Extension Service
College of Food and Natural Resources
Energy Conservation Program
Tillson Farm
Amherst, Massachusetts 01003
(413) 545-2132

Michigan

Michigan Energy Administration
6250 Mercantile Way, Suite 15
Lansing, Michigan 48913
(517) 374-9090

Michigan Solar Energy Association
201 E. Liberty St., Suite 2
Ann Arbor, Michigan 48104
(313) 663-7799

Minnesota

Minnesota Energy Agency
American Center Building
150 E. Kellogg Blvd.
St. Paul, Minnesota 55101
(612) 296-5120

Mississippi

Mississippi Energy Office
Suite 228, Bearfield Complex
455 N. Lamar
Jackson, Mississippi 39202
(601) 354-7406

Mississippi Solar Energy Association
c/o Dr. Pablo Okhuysen
225 W. Lampkin Rd.
Starkville Mississippi 39759
(601) 323-7246

Missouri

Division of Energy
P.O. Box 176, 1014 Madison St.
Jefferson City, Missouri 65102
(314) 751-4000; 1-800-392-0717
   (Missouri only)

Montana

Montana Department of
   Natural Resources
and Conservation
Energy Division
32 S. Ewing St.
Helena, Montana 59601
(406) 499-3940

Nebraska

Nebraska Energy Office, State Capitol
P.O. Box 95085
Lincoln, Nebraska 68509
(402) 471-2867

Nebraska Solar Energy Association
c/o Dr. Bing Chen
University of Nebraska
Department of Electrical Technology
60th and Dodge St.
Omaha, Nebraska 68182
(402) 554-2769

Nevada

Nevada Department of Energy
1050 E. William, Suite 405
Carson City, Nevada 89701
(702) 885-5157

New Hampshire

Governor's Council on Energy
2½ Beacon St.
Concord, New Hampshire 03301
(603) 271-2711

New Hampshire Solar Energy
   Association
P.O. Box 666
Manchester, New Hampshire 03105
(603) 435-8157

Northern New Hampshire Solar Energy
   Association
c/o Paul Hazelton
EVOG, Hebron, New Hampshire 03241
(603) 744-8918

New Jersey

New Jersey Department of Energy
Office of Alternate Technology
101 Commerce St.
Newark, New Jersey 07102
(201) 648-6293

New Mexico

New Mexico Energy and Minerals
   Department
P.O. Box 2770
Santa Fe, New Mexico 87503
(505) 827-2472

Alamogordo Solar Energy Association
c/o Ed Tyson
1832 Corte Del Ranchero
Alamogordo, New Mexico 88310
(505) 437-4258

Albuquerque Solar Energy Association
c/o Bob Stromberg
Solar Technical Division
Sandia Labs 4714
Albuquerque, New Mexico 87185
(505) 264-2282

Dona Ana Solar Energy Association
c/o Harry Zweibel
P.O. Box 1592
Las Cruces, New Mexico 88001
(505) 646-1846

New Mexico Solar Energy Association
P.O. Box 271
Santa Fe, New Mexico 87501
(505) 471-2573

San Miguel County Solar Energy
   Association
P.O. Box 153
Montezuma, New Mexico 87731

Taos Solar Energy Association
c/o Fred Hopman
P.O. Box 2334
Taos, New Mexico 87571
(505) 758-4051

New York

New York Energy Office
2 Rockefeller Plaza
Albany, New York 12223
(518) 473-8251

Eastern New York Solar Energy Society
P.O. Box 5181
Albany, New York 12205
(518) 270-6301

Metropolitan New York Solar Energy
   Association
c/o Mr. William Bobenhausen, president
P.O. Box 2147, Grand Central Station
New York, New York 10017
(914) 856-6633

North Carolina

North Carolina Energy Division
P.O. Box 25249
430 N. Salisbury
Raleigh, North Carolina 27611
(919) 733-2230

North Carolina Solar Energy
   Association
Suite 614, Tower 1
1110 Navoho Drive
Raleigh, North Carolina 27609

North Dakota

Federal Aid Coordinator
Office, Energy Management and
   Conservation
1533 N. 12th St.
Bismarck, North Dakota 58501
(701) 224-2250

Ohio

Ohio Department of Energy
30 E. Broad St.
Columbus, Ohio 43215
(614) 466-8277, 1-800-282-9234
   (Ohio only)

Oklahoma

Oklahoma Department of Energy
4400 N. Lincoln Blvd., Suite 251
Oklahoma City, Oklahoma 73105
(405) 521-3941

Oklahoma Solar Energy Association
c/o Dr. Bruce V. Ketcham
Solar Energy Laboratory
University of Tulsa
Tulsa, Oklahoma 74104
(918) 939-6351

Oklahoma Solar Energy Industries
  Association
4432 South 74th East Avenue
Tulsa, Oklahoma 74145

Oregon

Oregon Department of Energy
Labor and Industries Building
Room 102
Salem, Oregon 97310
(503) 378-6715

Columbia Solar Energy Association
4015 S.W. Canyon Rd.
Portland, Oregon 97221
(503) 242-0643

Pennsylvania

Governor's Energy Council
1625 N. Front St.
Harrisburg, Pennsylvania 17102
(717) 783-8610

Mid-Atlantic Solar Energy Association
2233 Grays Ferry
Philadelphia, Pennsylvania 19146
(215) 963-0880

Rhode Island

Governor's Energy Office
80 Dean St.
Providence, Rhode Island 02903
(401) 277-3374

South Carolina

Department of Energy Resources
Edgar Brown Building
1205 Pendleton St.
Columbia, South Carolina 29201
(803) 758-2050

South Dakota

South Dakota State Energy Office
Capital Lake Plaza
Pierre, South Dakota 57501
(605) 773-3604

Tennessee

Tennessee Energy Authority
226 Capitol Blvd., Suite 707
Nashville, Tennessee 37219
(615) 741-2994

Tennessee Solar Energy Association
P.O. Box 19
Middle Tennessee State University
Murfreesboro, Tennessee 37132
(615) 898-2778

Texas

Texas Energy and Natural Resources
  Advisory Council
411 W. 13th Street, Suite 804
Austin, Texas 78701
(512) 475-5407

Texas Solar Energy Association
c/o Russell E. Smith
1007 S. Congress, Suite 348
Austin, Texas 78704
(512) 443-2528

Utah

Utah Energy Office
231 E. 400 South, Suite 101
Salt Lake City, Utah 84111
(801) 533-5424;
Energy hotline: (801) 581-5424;
Toll-free in Utah: (800) 662-3633

Vermont

Vermont Energy Office
State Office Building
Montpelier, Vermont 05602
(802) 828-2393

New England Solar Energy Association
P.O. Box 541, 22 High St.
Brattleboro, Vermont 05301
(802) 254-2386

Virginia

Division of Energy
310 Turner Rd.
Richmond, Virginia 23225
(804) 745-3245

Virginia Solar Energy Association
c/o Larry Perry, P.O. Box 12442
Richmond, Virginia 23231
(703) 342-1816

Washington

Washington State Energy Office
400 E. Union
Olympia, Washington 98504
(206) 754-1350

Pacific Northwest Solar Energy
  Association
c/o Ecotope, 2332 E. Madison
Seattle, Washington 98112
(206) 322-3753

Western Sun, Washington Office
c/o SMT Program
318 Guggenheim, FS-15
University of Washington
Seattle Washington 98195
(206) 543-1249

Western Washington Solar Energy
  Association
c/o Ed Kennell
3534 Bagley N.
Seattle, Washington 98103
(206) 633-5505

West Virginia

West Virginia Fuel and Energy Office
1262½ Greenbrier St.
Charleston, West Virginia 25311
(304) 348-8860

Wisconsin

Wisconsin Division of State Energy
101 S. Webster Street
Madison, Wisconsin 53702
(608) 266-8234

Wisconsin Solar Energy Association
c/o Ernest Rogers
6704 Spring Grove Ct.
Middleton, Wisconsin 53562
(608) 831-4446

Wyoming

Wyoming Energy Conservation Office
320 W. 25th St.
Cheyenne, Wyoming 82002
(307) 777-7131

Index

Aeration of pond, 104, *illus.* 105
Aerodynamics of wind turbines, 8–10 and *illus.*
Airstream and rotor, 15, *illus.* 16
Alternating current (ac), 81–84, *illus.* 83; *see also* Inverters
Alternator, 82–84
American Wind Energy Association, 5, 7, 15 n.
Anemometer:
  hand-held, 37 and *illus.*
  wind-cup, 39 and *illus.*
Appliances:
  and calculating power needs, 50–54 and *tables*
  energy requirements of, *table* 48–49
  load cycle history of, 51–53 and *illus.*
Average wind power, *see* Power, wind: averages

Batteries, 86–89 and *illus.*
  bank, 87–88 and *illus.*, *illus.* 159
  capacity/number determination, 21–23, 87
  cost, 88–89
  and energy cycles, 21
  gassing of, 159–60
  installation of, 159–60 and *illus.*
  and water-pumping system, 62
  for wind-electric system, 84
Beaufort Scale, 3, 37, *table* 38
Blade:
  angle of, 78, *illus.* 79, 80 and *illus.*
  construction, *illus.* 76, 77
  icing of, 80 and *illus.*
  loads on, *illus.* 77
  maintenance of, 160
  number of, 71
Blockage of air flow, *see* Obstructions of air flow; Wind fence
Brakes, hydraulic, 80

Building codes, 117–19, 155
Building permit, 114–19
Buildings, effect on wind speed, 172, *illus.* 174
Buying wind systems, 7

Cantilever (freestanding) tower, 96–97 and *illus.*
Centrifugal force, and blades, 77
Climatological Data, 37–38
Consultants:
  on installation, 108
  meteorological, 43
  on tower design, 97
Cooperatives, 120
Cost:
  of batteries, 88–89
  investment, 110–11 and *table*
  of maintenance, 105–106
  of pond energy storage, 89
  of rated power (per kw), 77
  of system components, 107–108 and *illus.*
  of tower, 96, 169
  of wind survey, 41–42
Cost analysis, 6, 106–12 and *tables*
  comparing wind systems, 104–106 and *tables*
  and inflation, 111 and *tables*
  see also Depreciation; "First cost" factor
Cut-in speed, 74
Cut-out speed, 81

Darrieus wind turbine, 10, *illus.* 11, 12, *illus.* 80
  rotor, illus. 66, 77, *illus.* 80
  speed control of, 80–81 and *illus.*
Decisions about wind power, 2–7; *see also* Site selection; System: selection of

Density of air, 16–17
  k values, *table* 17
Depreciation, 105–106, 108, 110

Direct current (dc), 81–84, *illus.* 83; *see also* Inverters
Direction of wind, 39; *see also* Wind rose
Downwind rotor, *illus.* 72
DRA (Density Ratio at Altitude), *see* Density of air
Drag effect, 8–10 and *illus.*
  and towers, 95–96 and *illus.*
  and wind turbines, 10–12 and *illus.*
Drag spoiler, 80
DRT (Density Ratio at Temperature), *see* Density of air

Economic considerations, *see* Cost; Cost analysis
Efficiency, 15–17, 68, *illus.* 71
"Eggbeater" wind turbine, *see* Darrieus wind turbine
Electric energy demand curve, 61 and *illus.*, 62
Electric energy sales, 119–21
Electric generator, *see* Generator
Electric power, 12–13, 47
  wind generation of, *see* Electric system
Electric system, 6, 81–86 and *illus.*
  backup system, 84, 86 and *illus.*
  synchronous inverter type, 94–95 and *illus.*
  wiring, 84, *illus.* 85–86
  see also Energy storage; Generator; Wind furnace
Electric utilities, *see* Utilities
Energy:
  losses, 46–47 and *illus.*
  and power, 12–13
Energy cooperative, 120

Energy density, 88
Energy requirements:
    of appliances and farm equipment,
        table 48-49
    calculation of, 2, 6, 46-62 and *illus.*
    monthly, 54 and *illus.*
    steps, 50-54 and *illus.*
    energy demand curve, 61 and *illus.*,
        62
Energy rose, 21, *illus.* 22
Energy storage:
    with batteries, *see* Batteries
    hot air or water system, 91-93 and
        illus.
    pond system, 89-90 and *illus.*
    tank system, 60-62 and *illus.*
Enertech windmill, *illus.* 118, *illus.*
    128
Environmental considerations, 2, 116,
    160-61

Fan tail, 71, *illus.* 72
Farm equipment, energy requirements
    of, *table* 49
Farm windmill, *illus.* 5, 64-65 and
    illus., 66
Fastest mile, 165
Feathering blades, 78, *illus.* 79, 80
    and *illus.*
Federal Aviation Administration regu-
    lations, 117
"First cost" factor, 100, 104
Flagging (deformation of plants), 38
Flywheels, 93 and *illus.*
Furling speed, 81

Gear ratio, 81
Generator, 13, 81-86 and *illus.*
    auxiliary, 84, 86 and *illus.*, 99
    basic operation of, 67
    location on wind turbine, 81
    power take-off (PTO) type, 99
Gin pole, 155 and *illus.*
Ground roughness, 169, *illus.* 170,
    171 and *illus.*; *see also* Obstruc-
    tions to air flow
Guy-wire supported tower, 95-97 and
    illus., 161

Head loss factor, 55, *illus.* 56, 57
Heating system, *see* Wind furnace
Height and wind speed, 20, 28-30 and
    illus., 167, 169, *illus.* 170
Horizontal axis wind turbine, *illus.* 67
    rotors, *illus.* 69

Icing of blades, 80 and *illus.*
Inflation and cost of energy, 111 and
    tables
Installation:
    of tower, 155 and *illus.*
    of wind system, 108, 154-61 and
        illus.
    of wind turbine, 155, *illus.* 156, 157
        and *illus.*

Insurance, 110, 123, 154
Interference of air flow, *see* Obstruc-
    tions to air flow
Inverters, 5, 97-98 and *illus.*; *see also*
    Synchronous inverter

Lake for energy storage, 89-90 and
    illus.
Legal considerations, 2, 113-23
    liability, 7, 122
Lift, 9-10 and *illus.*
    and wind turbines, 10-12 and *illus.*
Lightning protection, 158-59 and *illus.*
Line loss, 82
Load cycle history of appliances,
    51-53 and *illus.*
Load monitor, 84, *illus.* 85, 86, 98

Maintenance and repair, 160
    cost, 105-106, 109
Manufacturers:
    liability of, 121-22
    standards, 15 n.
Mountain winds, 33-35 and *illus.*

National Climatic Center, Asheville,
    N.C.:
    Climatological Data, 37-38
    hourly record of weather, 43
    *Percentage Frequency of Wind Direc-
        tion and Speed*, 43, illus. 44, 167
    records of wind power averages, 26
National Weather Service, 24
"Negative easement," 114
Noise of wind system, 2, 161

Obstructions to air flow, 4, 172,
    174-75 and *illus.*
    and estimate of wind power, 36-37
    and placement of wind turbine, 35
    see also Buildings, effect on wind
        speed; Ground roughness

Panemone wind turbine, 9 and *illus.*
*Percentage Frequency of Wind Direc-
    tion and Speed*, 43, *illus.* 44, 167
Pond:
    for energy storage, 89-90 and *illus.*
    prevention of freezing, 104, *illus.* 105
Power, electric, 12-13, 47
    wind generation of, *see* Electric
        system
Power, mechanical, 12-13
Power, wind, 8-17 and *illus.*
    averages, 20, *illus.* 21, 26, *map* 27,
        28
        in U.S. and Canada, *tables* 137-53
        for wind duration curve, 74 n.
        for wind power estimate, 37, 41, 43,
            45, 175-77 and *tables*
        see also *Percentage Frequency of
            Wind Direction and Speed*
    calculating, 14-17, 24-26 and *illus.*
    curve, 72, *illus.* 73, *illus.* 74
    distribution, 26, *map* 27, 28

Power, wind (cont.)
    estimating, 36-39 and *illus.*, 175-77
        and *tables*
        survey (steps), 40-45, *illus.* 42
    local variations, 18-19 and *illus.*,
        30-35 and *illus.*; *see also* Ob-
        structions to air flow
    variations with time, 19-26 and *illus.*
    and wind speed, 13-14, 20
        calculating, 24-26 and *illus.*, 45
Power companies, *see* Utilities
Power load, 47
Power requirements, *see* Energy re-
    quirements
Propeller type wind turbine, 10, *illus.*
    11, *illus.* 66, 77-81 and *illus.*
    direction control, 71, *illus.* 72
Public utilities, *see* Utilities
Public Utility Regulatory Policies Act
    (1978), 94 n., 119 n.
Pumping water, *see* Water pumping

Recording devices, 40
Repair, *see* Maintenance and repair
Resale value of wind system, 109
Return time, 22
Ridges, effect on wind speed, 172,
    illus. 173
Rotor (windwheel), 63, 77-81 and
    illus.
    Darrieus type, *see* Darrieus wind
        turbine
    downwind type, *illus.* 72
    efficiency of, 15, 68, *illus.* 71
    Savonius type, *see* Savonius rotor
    selection of, 100
    speed control of, 77, 78 and *illus.*,
        illus. 79, 81
    upwind type, *illus.* 72
    see also Blade; Tip speed ratio
Rotor-swept area, *see* Swept area

Safety, 97, 122, 155, 159 n.
Sailwing blade, *illus.* 76
Savonius rotor, *illus.* 65, 66
    direction control of, 71
    ground control of, 80
    for pond aeration, 104, *illus.* 105
Sea breeze, *illus.* 19, 30, 31
    speed of, 31 and *illus.*
Selling wind energy, 119-21
Shelterbelt, 175
Sharing wind energy, 120
Silos and trees, and wind speed,
    174-75 and *illus.*
Site selection, 26-45, 166-77 and *illus.*
Site survey, 3-4, 40-45, *illus.* 42
    cost, 41-42
    and energy storage needs, 60
Small Wind Systems Test Center
    Rocky Flats, Colo., 5, 108, 165
Solidity, 66, 68 and *illus.*, 71
Spark arrester, 159 and *illus.*
Speed-cubed effect, 13-14 and *illus.*, 20
Stalling, 80
Storage batteries, *see* Batteries

Storm winds, 165, 167
Survey, *see* Site survey; Wind survey
Survival speed, 165
Swept area:
  and average annual energy, 25
  calculating, 66–67
  formula, 17 n.
  and solidity, 66
Synchronous generator, 82
Synchronous inverter, 5–6, 94–95 and
  illus.
  and selling wind power, 121
System, *illus.* 63
  components, 4–6, 63–99 and *illus.*,
    illus. 102, 103
    cost, 107–108
    see also individual components
  cost, *see* Cost analysis
  efficiency of, 15–17
  installation of, 108, 154–61 and *illus.*
  purchasing, 7
  resale value of, 109
  selection of, 100–112 and *illus.*
    examples, 124–31
  for water pumping, *see* Water pump-
    ing system
  see also Electric system; Turbine;
    Wind furnace

Taxes, 109–10
Terrain, *see* Ground, roughness of
Testing wind turbines, 5, 108, 165, 167
Tip speed ratio, 67–68
Torque, 67, 68 and *illus.*
Tower, 95–97 and *illus.*
  installation of, 155 and *illus.*
Transformer, 82–83 and *illus.*
Trees:
  deformed by wind, 38
  measuring height of, 36–37 and *illus.*
  as obstruction to air flow, 36–37
  and wind speed, 174–75 and *table*
Turbine:
  aerodynamics of, 8–10 and *illus.*
  cut-in speed of, 74
  efficiency, 68, *illus.* 71
  horizontal-axis type, *illus.* 67
    rotors, *illus.* 69
  installation of, 155, illus. 156, 157
    and *illus.*
  and lift/drag, 10–12 and *illus.*
  maintenance of, 160

  manufacturer's standards, 15 n.
  power output of, 72–77 and *illus.*

Turbine (cont.)
  propeller-type, *see* Propeller-type
    wind turbine
  rated wind speed of, 74, 77
  size, 4
  speed control, 77, 78 and *illus.*,
    illus. 79, 81
  survival speed of, 165
  testing, 5, 108, 165
  types, 5, 68, *illus.* 69, *illus.* 70
  vertical-axis type, *see* Vertical-axis
    wind turbine
  see also Blade; Rotor
Turbulence:
  and buildings, 172
  and mountain winds, 34
  and obstructions to air flow, 35
  and ridges, 172
  and trees, 174, *table* 175
  and valley winds, 32

Upwind rotor, *illus.* 72
Utilities (power companies):
  rates, inflation of, 111
  and sale/purchase of private wind
    power, 119–21
  and synchronous inverter, 94–95

Valley winds, 32–33 and *illus.*
Velocity-cubed effect, 13–14 and *illus.*,
  20
Vertical-axis wind turbine, *illus.* 67
  ground control of, 80
  rotors, *illus.* 70

Warranty, 121–22
Water head, 54–55
  head loss factor, 55, *illus.* 56, 57
Water heating, 91–93 and *illus.*
Water pumping system, 2–3, *illus.* 8,
  illus. 55, *illus.* 64, *illus.* 103
  components, 6
  electric-energy demand curve, 61 and
    illus., 62
  estimating mechanical load, 54–60
    and *illus.*
  selection of, 103–104 and *table*
    examples, 124–31
Water requirements, 58, *table* 59
  for energy storage, 89, *illus.* 90
WBAN 10A (form), 43, 175
WECS (Wind energy conversion system), *see* System

Wind circulation, worldwide, *map* 18
Wind-cup anemometer, 39 and *illus.*
Wind direction, 39; *see also* Wind rose
Wind duration curve, 25–26 and *illus.*,
  43, 74–75 and *illus.*
Wind-electric system, *see* Electric
  system
Wind fence, 78, *illus.* 79
Wind furnace, 5–6, 91–93 and *illus.*
Wind machines, *see* Turbines
Wind measurement, *see* Wind direction;
  Wind speed: measurement of
Windmill, *see* Turbine
Wind power, *see* Power, wind
"Wind rights," 2, 114
Wind rose, 21, *illus.* 22, 43
  for U.S.A., 167, *map* 168
Wind speed:
  averages, 20, *illus.* 21, *illus.* 24;
    see also Power, wind: averages
  calculation of, 4
  and height, 20, 28–30 and *illus.*, 167,
    169, *illus.* 170
  maximum, *table* 166, 167
  measurement of, 37 and *illus.*, 39–40
    and *illus.*
  profile, 28–30 and *illus.* 167, 169
  and ridges, 172, *illus.* 173
  of sea breeze, 31 and *illus.*
  survival speed, 165
  and terrain, 169, *illus.* 170, 171 and
    illus.
  variations with time, 23 and *illus.*,
    165, *illus.* 166
  and wind power, 13–14, 20, 24–26 and
    illus., 45
Wind survey, 40–45, *illus.* 42, 60
  cost of, 41–42
Wind system, *see* System
Wind turbine, *see* Turbine
Wind turbulence, *see* Turbulence
Windwheel, *see* Rotor
Wiring, 157–58 and *illus.*
  for battery bank, 87, 88 and *illus.*
  for heating systems, *illus.* 92
  for wind-electric system, 84, *illus.*
    85–86
WTG (wind turbine generator), *see*
  Generator

Yaw, 71, 72

Zoning ordinances, 2, 114–16

Other Garden Way Books You Will Enjoy